AF541570

Insect Taxonomy

About the Authors

Dr. Bikash Subba is an Assistant Professor of Agricultural Entomology at the School of Agricultural Science, GD Goenka University. His academic journey includes previous roles as a lecturer at Sikkim University and ITM University. Dr. Subba's expertise spans several critical areas within entomology, including Biological Control of Pests and Diseases, Insect Ecology, Insect Taxonomy, Integrated Pest Management, Crop Protection Entomology, and Pollination, with a particular focus on insect pollinators.

Dr. Subba is a distinguished entomologist with a strong academic foundation. Having earned his MSc and PhD in Agricultural Entomology from Uttar Banga Krishi Viswavidyalaya in 2013 and 2019, respectively, he has consistently demonstrated a commitment to advancing knowledge in his field. His research prowess is evident in his publication of 15 research papers in esteemed journals and his contributions to Several book chapter. Moreover, Dr. Subba's active participation in national and international conferences underscores his dedication to sharing his findings and engaging with the broader scientific community.

Beyond his scholarly contributions, Dr. Subba is deeply committed to translating his research into tangible benefits for society. His dedication to practical applications is exemplified by his numerous workshops aimed at equipping farmers with organic farming techniques. By bridging the gap between academia and agriculture, Dr. Subba demonstrates a remarkable ability to foster sustainable practices and empower rural communities. His work underscores the importance of a holistic approach to entomology, one that seamlessly integrates rigorous research with real-world applications.

Dr. Biwash Gurung is currently working as an Assistant Professor in School of Agricultural Sciences (ICAR Accredited) G.D Goenka University Gurugram, Haryana. He has obtained his Ph.D Degree in Agricultural Entomology from Uttar Banga Krishi Viswavidyalaya, (North Bengal Agricultural University), West Bengal with prestigious University Research Scholarship (URS) sponsored by Government of West Bengal in the year 2017 for the period of 4 years. He has qualified ICAR -NET 2018. He holds one year eight months of teaching experience at ITM University, Gwalior, before joining G.D Goenka University. He has published more than 18 research paper, 2 short communications, 2 Book chapters, 6 popular articles, 1 paper at Book of proceedings and 1practical manual book. Dr. Gurung has reported "New prey record of giant ladybird beetle Anisolemnia dilatata (Fabricius) (Coccinellidae: Coleoptera feeding on Som plant Aphid Aiceona sp". Dr. Gurung has delivered Plenary lecture at 11th International Conference on Technology, Innovation, and Management for Sustainable Development (TIMS- 2023). He has also attended workshops, conferences and symposium at National and International level.

Insect Taxonomy
An Illustrated Guide

Bikash Subba
Biwash Gurung

1168, Sector 13, Urban Estate
Karnal-132 001, Haryana
Tel: 91-84470 75807, 18440 41168
Email: contentvibesppa@gmail.com
www.contentvibes.in

Print ISBN 978-93-6755-003-8

eISBN: 978-93-6755-205-6

Foreword

Introducing this on Insect Taxonomy: An Illustrated Guide authored by Dr. Bikash Subba and Dr. Biwash Gurung fills me with great enthusiasm. Their deep expertise and meticulous attention to detail have produced Comprehensive Guide of Insect Taxonomy that is an invaluable resource for both seasoned researchers and budding enthusiasts.

Insect taxonomy, as elucidated in this Insect Taxonomy: An Illustrated Guide, serves as a foundation in our understanding of Earth's biodiversity. With over a million documented species, insects are pivotal to life on our planet, driving processes like pollination, nutrient cycling, and pest control. By organizing these creatures into a hierarchical system, taxonomy provides a framework for scientific exploration across countless disciplines.

From ecology to agriculture, medicine to conservation biology, taxonomy offers essential tools for researchers to discern ecological roles, trace evolutionary relationships, and predict how organisms may respond to environmental changes. This understanding is pivotal in informing conservation strategies, managing habitats effectively, and developing sustainable practices that preserve biodiversity while addressing challenges in pest management and disease control.

Dr. Subba and Dr. Gurung's Insect Taxonomy: An Illustrated Guide extends beyond theoretical knowledge by imparting practical skills in insect identification using dichotomous keys. Through detailed descriptions and illustrations, readers are equipped to master the nuances of identifying key anatomical features, particularly within the Order and its suborders and families. This expertise not only enhances academic pursuits but also accentuates the practical significance of accurate insect identification in fields ranging from agriculture to environmental monitoring.

Commencing on this entomological exploration through the pages of this Guide, you will not only deepen your understanding of insect taxonomy but also cultivate a profound appreciation for the vital roles insects play in sustaining our ecosystems. Whether you are just beginning to explore entomology or refining your expertise as a proficient expert, Dr. Subba and Dr. Gurung's guide will broaden your knowledge and passion for the fascinating world of insects.

Readers are invited to immerse themselves in this comprehensive resource, where each chapter encourages exploration, learning, and discovery of the elaborate details of insect life through taxonomy. May this journey ignite curiosity, broaden perspectives, and inspire fascination with these remarkable creatures contributing to our planet's biodiversity and ecological harmony.

Prof. (Dr.) Shiv Sing Tomar
Dean, School of Agricultural Sciences
GD Goenka University
Sohna-Gurgaon Road, Haryana, India.

Preface

The study of insects is foundational to biology, offering profound insights into their diverse forms and ecological significance. This comprehensive guide serves as a composite resource compiled to introduce students to the intricate art of insect identification using taxonomic keys. Accurate identification of insects is essential across various disciplines, including agriculture, ecology, and conservation biology.

Insects, classified within the class Insecta of the phylum Arthropoda, are characterized by segmented bodies, six legs, and typically one or two pairs of wings. However, distinguishing between different arthropod groups can be challenging without systematic tools such as dichotomous keys. These keys present paired statements that systematically guide observers through choices based on visible traits, enabling precise identification of specific insect orders or families to species.

This comprehensive guide is designed to enhance the proficiency of students and professionals in utilizing dichotomous keys effectively. Practical activities provided offer hands-on experiences aimed at refining students' abilities to recognize key anatomical features such as wing structure, antennae shape, and leg arrangement. By engaging directly with detailed descriptions and illustrations, students will develop a structured approach to insect identification, with a specific focus on the taxonomic characteristics of the Order and its subdivisions into suborders and families.

Furthermore, the manual emphasizes the practical implications of accurate insect identification. It guide underscores the importance of distinguishing between beneficial and harmful insect species to implement effective integrated pest management strategies in agriculture. Additionally, identifying indicator species is crucial for assessing environmental health and biodiversity in ecological studies.

Through these practical exercises, students not only hone their taxonomic skills but also gain a deeper appreciation for the ecological roles and evolutionary adaptations of insects. Mastery of insect identification using dichotomous keys forms a foundational skill set for aspiring biologists and environmental scientists.

We aim for this guide to serve as a comprehensive guide aimed at helping students excel in insect identification through practical, hands-on activities. Our hope is that these exercises will foster a lifelong curiosity and appreciation for the intricate world of insects and their indispensable contributions to our planet's ecosystems.

Authors

Contents

1

Insect Identification Using Taxonomic Keys

Introduction

Insects, classified under the Class Insecta within the Phylum Arthropoda, exhibit diverse characteristics. Sometimes, distinguishing between different arthropods is necessary to confirm their classification as insects. Biologists frequently employ dichotomous keys, systematic tools designed to assist in organism identification, to navigate this task.

Instructions

Follow the steps outlined below to identify the insects described. Utilize the provided dichotomous keys and accompanying figures to aid in the identification process.

Fig. 1

1. With 1-2 pairs of clear, visible, usually transparent wings (Fig. 1). (If it is not a bird or a bat, then it is an insect) ....Insecta
   - Without obvious wings.......................2

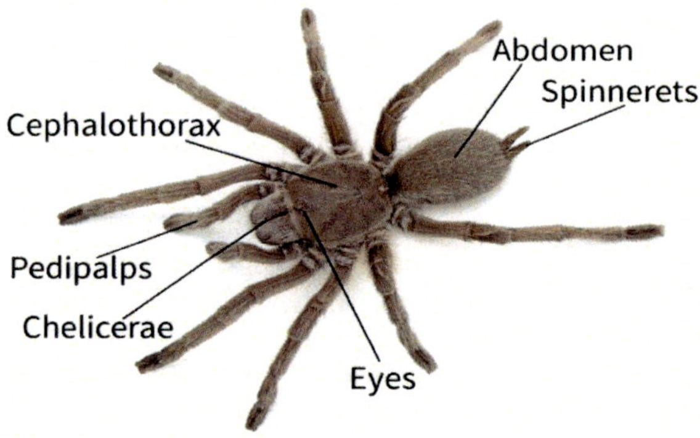

Fig. 2

2. With one or two pairs of antennae (segmented feelers) of varying shapes (see Figures 1, 3-5) located on the head, typically between the eyes. These antennae can be subtle, hidden beneath the head when viewed from above, or short and bristle-like. Caution: Some non-insect creatures might have front legs or modified mouthparts (pedipalps, Figure 2) that resemble antennae. 3
   - Lacking segmented antennae and always lacking any indication of wings.......................8.
3. With two pairs of antennae (one pair may be smaller than the other; the second pair is vestigial in terrestrial Isopoda such as pillbugs and sowbugs); body typically divided into two distinct regions (see Figures 3-4), the cephalothorax and abdomen; the number of legs on the cephalothorax varies, and the abdomen may or may not have appendages, which, when present, are not leg-like (examples include amphipods, sowbugs, lobsters, crayfish) ....Crustacea

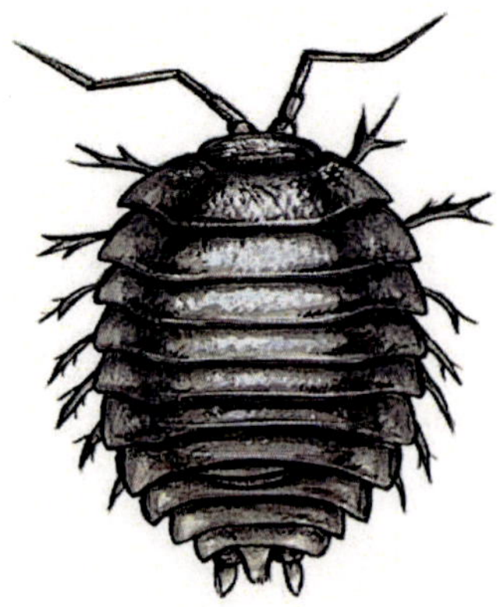

Fig. 3

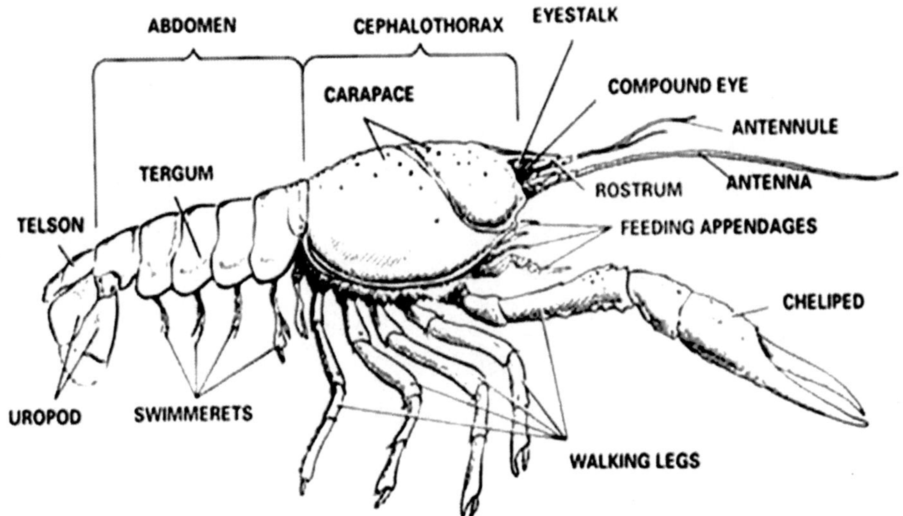

Fig. 4

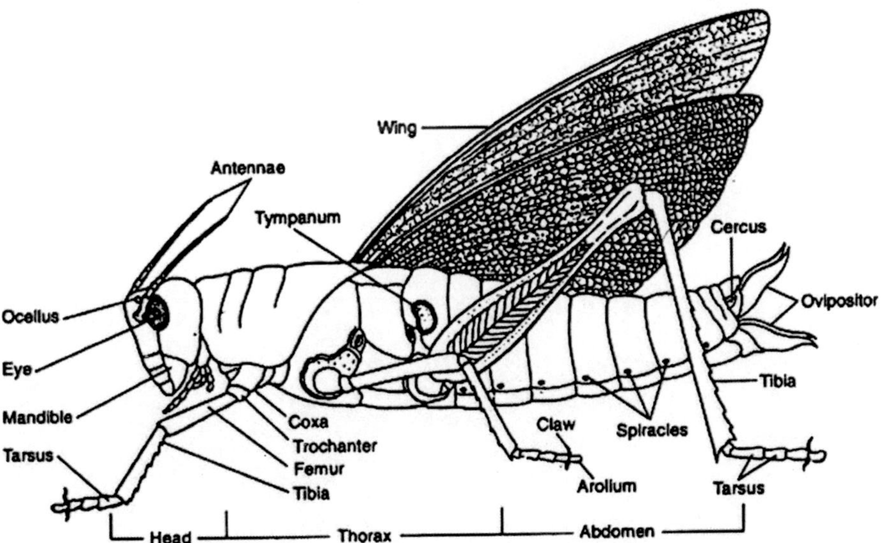

Fig. 5

4. With only 3 pairs of legs and often with 1-2 pairs of wings; body typically divided into 3 distinct regions (Fig. 5). Abdomen lacks segmented legs but may have appendages; body shape varies......................Insecta.

- With 9 or more pairs of legs (see Figures 6-7), mostly located on segments posterior to the head; distinct head; wings absent; body elongated and wormlike......................5.

5. Legs are evenly spaced along the body, typically with 1 pair of legs per segment....................... 6.
   - Legs arranged in pairs, with 2 pairs per segment (refer to Figure 6) belong to Diplopoda (millipedes)
6. Body is flattened with typically more than 15 pairs of legs, usually exceeding 25mm in size (see Figure 7a) (centipedes) .... Chilopoda
   - Body cylindrical; small forms with 9-12 pairs of legs 7.
7. Antennae branched (see Figure 7b); 9 pairs of legs belong to Pauropoda.
   - Antennae not branched; 10-12 pairs of legs (see Figure 7c) belong to Symphyla.

Fig. 6

Fig. 7

Fig. 7b: Pauropod

Fig. 7c: Symphylan

8. Typically possessing 7 pairs of appendages, including 5 pairs of legs, and exclusively inhabiting marine environments with a rudimentary abdomen.......................Pycnogonida.
   - Six pairs of appendages, occasionally fewer, typically with 4 to 5 pairs of legs; featuring a well-developed abdomen.......9

9. These are large marine species, reaching lengths up to 460mm, featuring an oval body protected by a hard shell and characterized by a long, spine-like tail (such as Horseshoe crabs, see Figure 7d...................... Xiphosura.

   - Smaller members, measuring less than 75mm, lack a hard shell and spine-like tail; this group encompasses spiders, ticks, mites, whip scorpions, windscorpions, and scorpions, ...Arachnida.

Fig. 7. d. Horse Shoe crab

Fig. 8: Hemelytra

Fig. 9: Tegmen

Example of Arachnids

An insect key is a tool designed for identifying different insect species. These keys are usually organized into pairs of choices called couplets. Each couplet offers two options based on specific features like the insect's size or the shape of its antennae. The user picks the option that most closely matches the insect they are examining. This choice then directs them to the next couplet. This process is repeated until the user reaches the final couplet, which reveals the identity of the insect.

Crab
Scorpion
Spider
Mites

1. Wings present (wings may be concealed under external elytra (pages 10-11), hemelytra (Figure 8), or tegmina (Figure 9) such that "wings" do not appear to be present) ....22.
   - Wings absent or reduced to small pads; many abdominal segments visible from above... Proceed to step 2.
2. Antennae absent; body slender and whitish in color. Very small (Figure 10) (1mm)............Protura.
   - Antennae present (may be difficult to see) ....3.
3. Usually equipped with a forked spring (furcula - see Figure 11) on the abdomen. Typically small, measuring 2-4mm, and always lacking apical abdominal cerci. In cases where the furcula is absent, size and body shape define the order Collembola.
   - Furcula always absent. Larger body size, various shapes.

Fig. 10: Protura **Fig. 11:** Collembola **Fig. 12:** Diplura

Fig. 13: Thysanura

Fig. 14: Homoptera

Fig. 15 Embiidina

Fig. 16: Thysanoptera

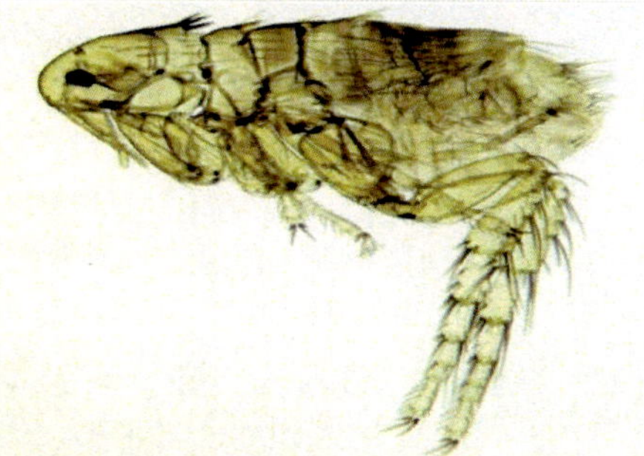

Fig. 16: Siphonaptera

1. Apex of abdomen with long cerci (Fig. 13) and lacking ventral abdominal styliform appendages or if ventral styliform appendages present, cerci are short
   - If cerci are short or absent, and abdominal styliform appendages are always absent.......................6.
2. Apex of abdomen with 3 filamentous cerci (silverfish) .... Thysanura.
   - Apex of abdomen with 2 cerci, either forceps-like (Fig. 12) or short and segmented ....Diplura
3. Large unsegmented forceps-like structures at apex of abdomen (earwigs) ....Dermaptera.
4. Cerci (when present) neither forceps-like nor unsegmented...................7.
5. Large insects, usually > 25mm in length; antennae frequently very long and slender.......................8
   - Small insects, usually < 12mm in length......................9.
6. If an insect has 4-segmented tarsi, it is classified under Orthoptera.............5
   - segmented tarsi ....Phasmida

7. Tube-like structures (cornicles) extending from the 4th to the last abdominal segment, or a body covered with waxy filaments or a scale.......................Homoptera.

    - If an insect does not have cornicles and lacks a covering of scale or waxy filaments on its body......................10.

8. If an insect's abdomen is noticeably narrowed at the junction with the thorax, resembling a waist of bees, wasps, ants, sawflies...................... Hymenoptera.

    - If an insect's abdomen does not show a distinct narrowing at the junction with the thorax.........11

9. Mouthparts (rasping-sucking) contained in a short, cone-like beak; wings when present often with fringe of hairs (Fig. 16); size ; size <3mm; abdomen often pointed at apex (thrips) ....Thysanoptera

    - Mouthparts other than rasping-sucking may be in the form of an elongate beak that extends ventrally and posteriorly beneath head; 13

10. Insects characterized by a laterally flattened body adorned with numerous backward-projecting spines and bristles, along with long legs featuring greatly enlarged coxae adapted for jumping (Fig. 17) .... Siphonaptera

    - If an insect's body is not flattened laterally and may have hairs or spines that are not projecting backward, and if its legs are modified for jumping with enlarged femora......................14

11. Mouthparts elongated into piercing-sucking beak 15

    - Mouthparts not elongated into long piercing beak; head may be prolonged 16

12. Antennae hidden in grooves in head ....Diptera

    - Antennae long and easily seen (Fig. 18) Heteroptera

13. Body covered with dense hair ....Lepidoptera

    - Body lacking dense hair 17

14. Antennae moniliform (segments beadlike); short cerci present (Fig. 19) (termites) ....Isoptera

    - Antennae not moniliform; cerci absent 18

15. Antennae long and slender 19

- Antennae short 20

16. Head prolonged and beak-like (Fig. 20); males of some species have scorpion-like abdomen (scorpionflies.................Mecoptera

- Head not prolonged and beak-like ....Psocoptera

17. Tarsi with 4-5 segments ....Diptera

- Tarsi with 1-3 segment (lice) - Phthiraptera 21

18. Chewing mouthparts; head usually broader than long (Fig. 21a) ....Mallophaga

- Piercing-sucking mouthparts retracted into head; head usually longer than broad; legs greatly enlarged for grasping (body lice) (Fig. 21b) ....Anoplura

19. Abdomen with large unsegmented forceps-like cerci (Fig. 24)................... Dermaptera

- Cerci appearing segmented when present, not forcepslike, or absent...........................23

20. Cerci filamentous, longer than last 3 abdominal segments combined 24

- Cerci shorter than last 3 abdominal segments combined, not filamentous, or totally absent. 28

21. Wings folded upright and parallel to body length; antennae setaceous (Fig. 22. mayflies) ....Ephemeroptera

- Wings various but not held upright above body; antennae elongate and filiform 25

22. Front pair of legs shaped differently than mid and hind pair, modified for digging (Fig. 23a) (fossorial) or grasping (Fig. 23b) (raptorial) .... Orthoptera

- Front pair of legs similiar to middle pair 26

23. Hind pairs of legs enlarged for jumping (Fig. 23c)............................. Orthoptera

- Hind pair of legs similar to middle pair 2

Fig. 18. Heteroptera **Fig. 19:** Isoptera **Fig. 20:** Mecoptera

Fig. 21: Phthiraptera **Fig. 22:** Ephemeroptera

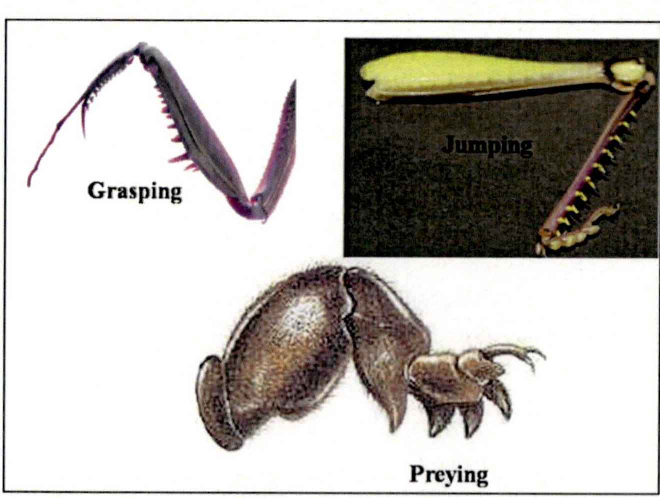

Fig. 23: Orthoptera types of legs

Fig. 24: Dermaptera

Fig. 25: Odonata wings

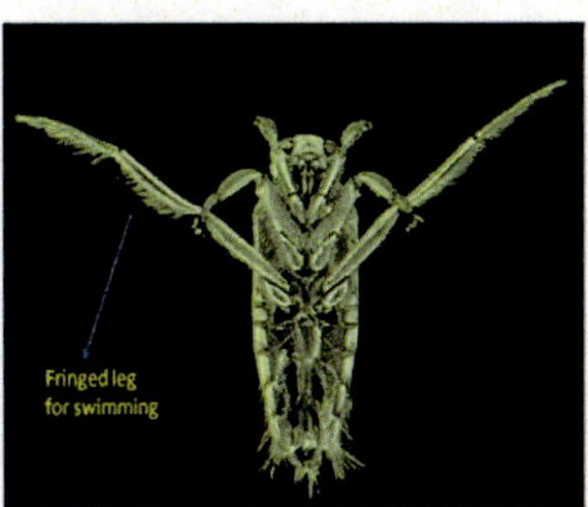

Fig. 26: Aquatic heteroptera

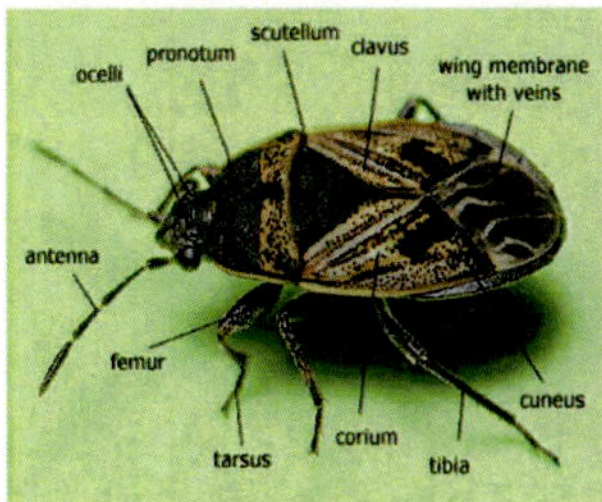

Fig. 27: Heteroptera, Hemi-elytra

Fig. 28: Coleoptera, forewings elytron

Fig. 29: Orthoptera (forewings, Tegmen

24. Insects with 3-segmented tarsi and cerci that are either long or short, not forceps-like, and typically many-segmented..................... Plecoptera.
    - Insects with variable tarsi (4-5 segments), often large and bulky, with well-developed wings.............. 31
25. Insects with cerci present, which are shorter than the last 3 abdominal segments combined....................29.
    - If an insect lacks cerci (without confusion from genitalia)............33
26. Small and delicate insects with transparent wings of uniform shape and size............30.
    - Insects with a varied body shape and wings that are in the form of elytra, tegmina, or hemelytra........ 31.
27. Insects with front basitarsi (1st tarsomere) enlarged and dilated to form a webspinning organ................Embiidina.
    - Front basitarsi not enlarged and dilated, appearing of normal proportions (termites) ....Isoptera
28. Insects with 4-segmented tarsi...................Orthoptera.
    - Tarsi 5 segmented 32

29. If the prothorax is much longer than the mesothorax and the front legs are modified for grasping..........Mantodea (praying mantises).
    - If the prothorax is not greatly lengthened and the front legs are not modified for grasping..............Blattaria (cockroaches).
30. Large insects with 2 pairs of wings; wings usually transparent, each wing with an anterior node (Fig. 25)or notch (dragonflies, damselflies) ....Odonata
    - Wings that vary in appearance but do not have an anterior node or notch...................34
31. One pair of wings and halteres present...................Diptera.
    - two pairs of wings; halteres absent 35
32. If an insect's mouthparts are structured as a piercing-sucking, elongate beak that is mostly held beneath and behind the head, and if it lacks palpi......................36
    - Mouthparts other than above; palpi present 38
33. Hind leg lacks tarsal claws and is adapted for swimming (as depicted in Fig. 26) ....Heteroptera.
    - Hind leg with tarsal claws 37
34. Beak arises from anterior part of head; forewings usually as hemelytra (Fig. 27) ....Heteroptera
    - Beak appears to originate from between the front pair of legs and its forewings have a uniform texture..................Homoptera.
35. Insects with rasping-sucking mouthparts in the form of a cone-like beak and wings that are fringed with long hairs...................Thysanoptera.
    - Not as above 39
36. Insects with a front pair of wings that are hardened and of a different texture than the rear flight wings..................................40
    - Front wings of an insect are not thickened or hardened to form a cover for the flight wings 41
37. Front pair of wings is thickened and usually hard, without crossveins, meeting along the midline (meson) of the body to form elytra (as depicted in Fig. 28), and there are many forms with elytra shortened to expose one or more abdominal segments from above, hind legs usually not modified for jumping..................................Coleoptera (beetles).

- Front pair of wings has obvious crossveins and veins (as seen in Fig. 29, tegmen), and they overlap one another at least partially, and the hind legs are often enlarged for jumping............................ Orthoptera (grasshoppers, crickets, katydids)

38. If the front basitarsi (1st segment) of an insect are enlarged to form silk-producing glands (as depicted in Fig. 15) (webspinners)................. Embiidina

    - If the front basitarsi are not enlarged more than the remaining segments..............42

39. All wings equal in size; (termites) ....Isoptera

    - Hind wings usually smaller than front pair of wings; 43

40. Insects with mouthparts shaped like a coiled siphon (as depicted in Fig. 30) and with wings and body typically covered with scales (butterflies and moths).........Lepidoptera

    - Mouthparts are not in the form of a coiled siphon and its body lacks scales or has only a few, which are restricted to the wings and wing veins..................44.

41. If an insect's wings have many crossveins, especially at the anterior edge (as depicted in Fig. 31)................Neuroptera.

    - If an insect has few crossveins in its wings and lacks a waxy coating on its body and wings, proceed to step 45 for further classification.

42. Reduced, vestigial mouth and only obvious palpi, with hairs often present on the wings, ...Trichoptera (caddisflies).

    - Mouthparts not reduced or vestigial; chewing or chewing-lapping types 46

43. Chewing mouthparts elongated into a beak-like structure and some males exhibit a scorpion-like abdomen...............Mecoptera (scorpionflies).

    - If an insect does not have chewing mouthparts elongated into a beak or has chewing-lapping mouthparts.....................47.

44. Tarsi that are 4- or 5-segmented and wings that are folded flat over the body (as depicted in Fig. 32)..........Hymenoptera (bees, wasps, ants, and sawflies).

    - Tarsi that are 2- or 3-segmented and wings that are folded roof-like over the body (as depicted in Fig. 33), it belongs to the order Psocoptera, which includes booklice and barklice.

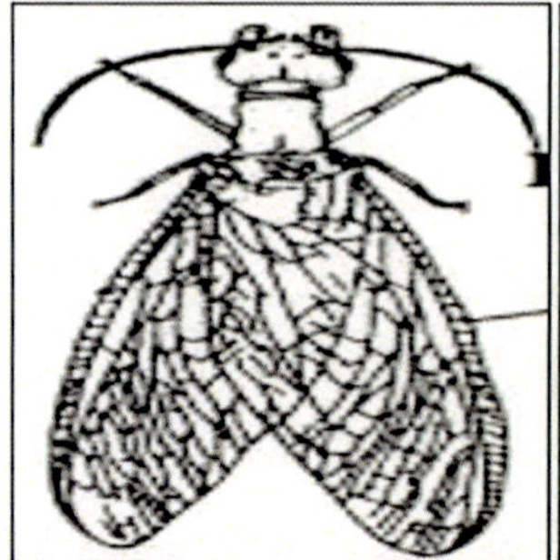

Fig. 30: Neuropteta including Megaloptera Cross veins

Fig. 31: Hymenoptera

Fig. 33: Psocoptera

Stag beetle (Cleoptera Lucanidiae) with elytra closed appearing to black typical wings

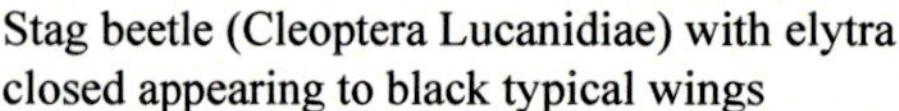

Stag beetle (Cleoptera Lucanidiae) with elytra closed appearing to black typical wings

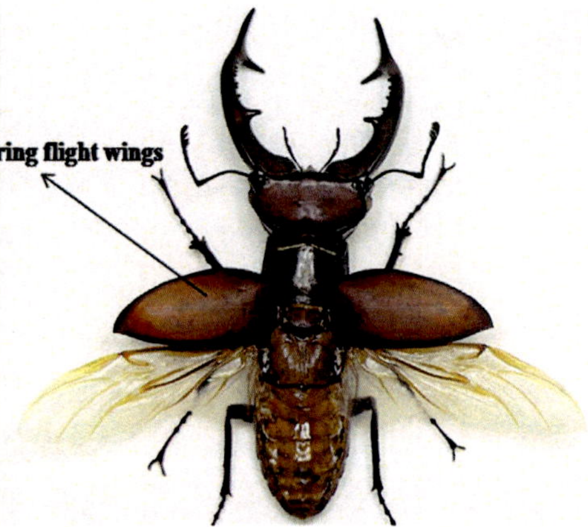

Stag beetle (Cleoptera Lucanidiae) with elytra open preparing to take flight, visible flight wings

Feathery antenna of male moths (Lepideoptera)

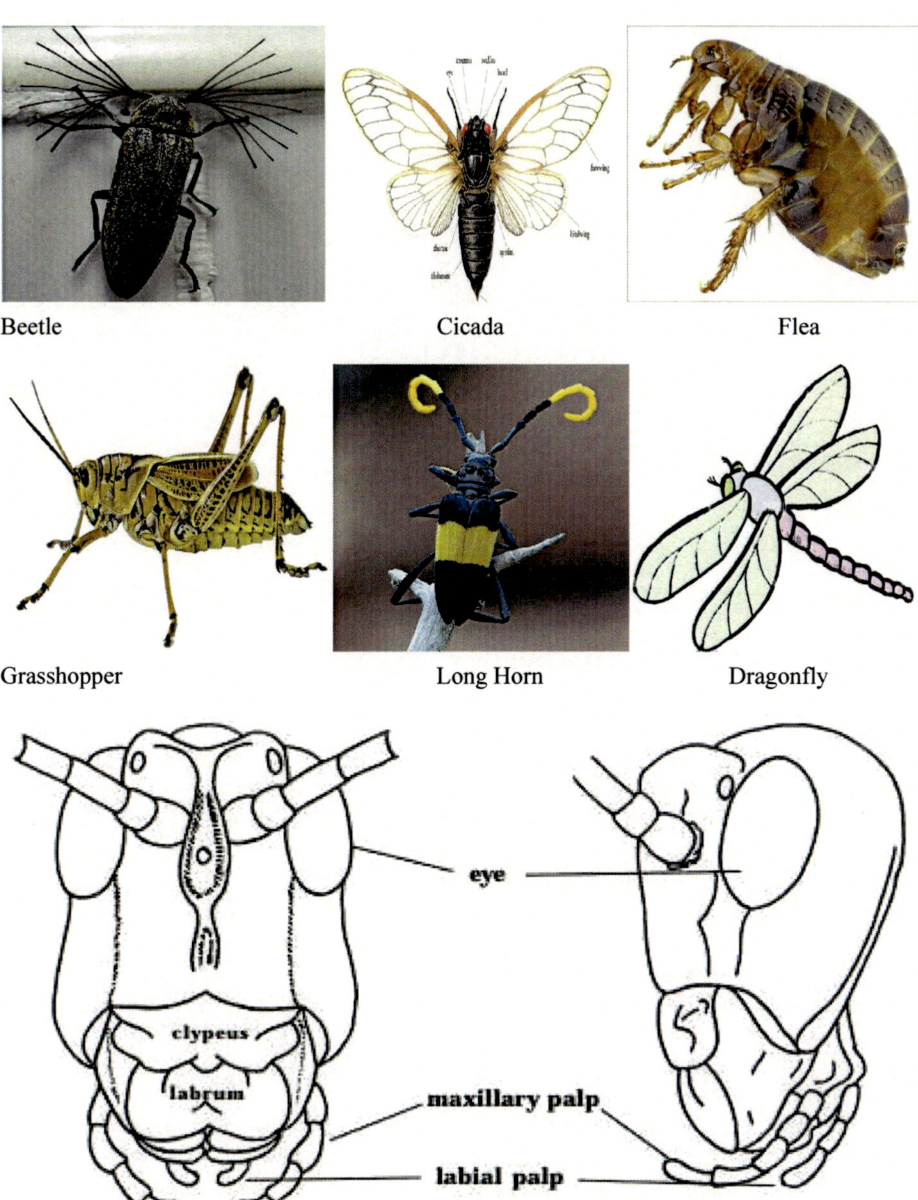

Beetle Cicada Flea

Grasshopper Long Horn Dragonfly

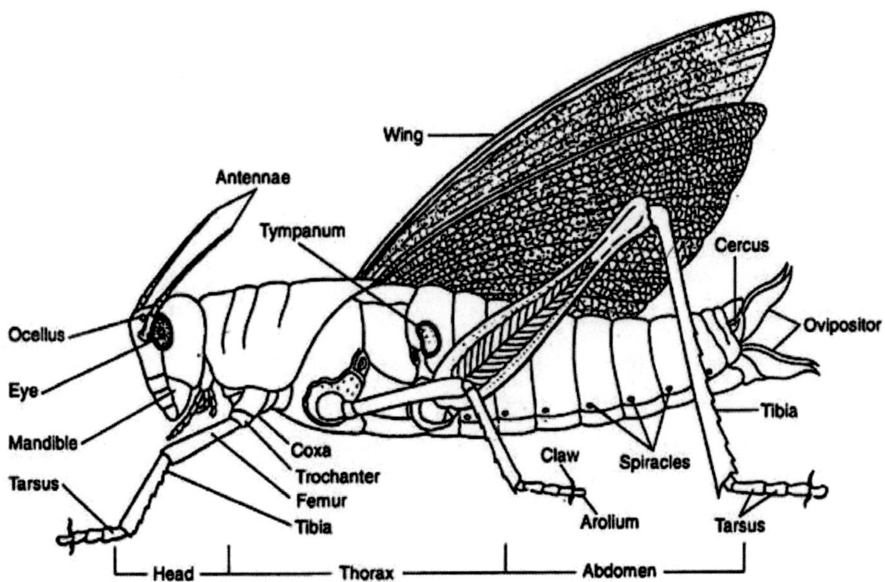

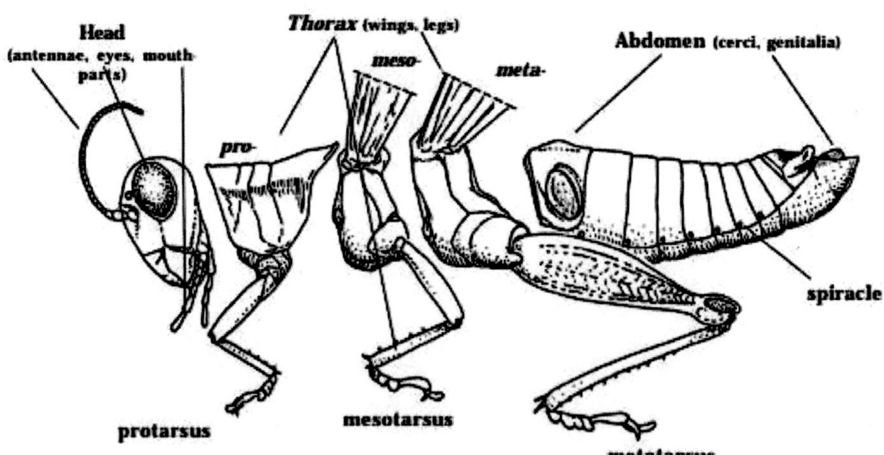

Body regions of grasshopper

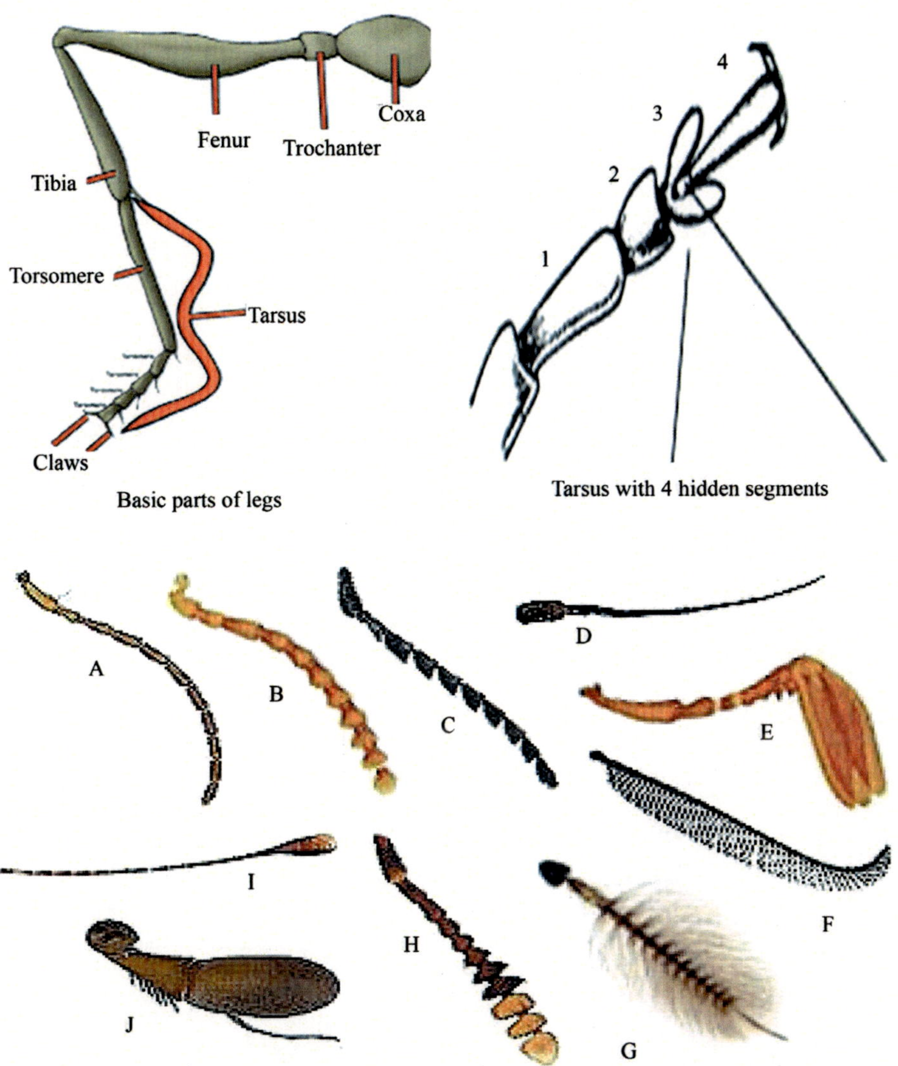

A: Filiform = thread-like, B; Moniliform = beaded, C; Serrate = sawtoothed, D; Setaceous = bristle-like, E; Lamellate = nested plates, F; Pectinate = comb-like, G; Plumose = long hairs, H; Clavate = gradually clubbed, I; Capitate = abruptly clubbed, J; Aristate = pouch-like with one lateral bristle

Activity

1. To study about the lepidopteran insect based on key.
2. To study about the coleopteran insect based on key

2

Insect Classification Based on Wing Structure

Objective

The aim of this practical is to categorize insects based on their wing characteristics.

Introduction: Insect classification, as outlined by A.D. Imms (1957), divides Class Insecta or Hexapoda into two primary groups:

Group 1: Apterygota: These insects lack wings, and the absence of wings is considered primitive. Metamorphosis is minimal or absent. Adults typically possess one or more pairs of pregenital abdominal appendages. The mandible of adults usually articulates with the head capsule at a single point. This group comprises four orders:

1. Thysanura - Example: Silverfish
2. Diplura - Example: Japygids
3. Protura - Example: Telson tails or proturans
4. Collembola - Example: Springtails

Group 2: Pterygota: These insects possess wings or may have lost them secondarily. Metamorphosis varies, and adults do not have pregenital abdominal appendages. The mandible of adults typically articulates with the head capsule at two points. This group is subdivided into two divisions:

Division I: Exopterygota: Wings develop externally. Metamorphosis is simple, with the pupal stage rarely occurring. Immature stages, known as nymphs, closely resemble adults in structure and behavior. This division includes various orders, such as:

- Ephemeroptera (Mayflies)
- Odonata (Dragonflies)
- Plecoptera (Stoneflies)

- Orthoptera (Grasshoppers & crickets)
- Phasmida (Phasmids)
- and others.

Division II: Endopterygota: These insects develop wings internally. Metamorphosis is complex, involving a distinct pupal stage. Immature stages, known as larvae, differ significantly from adults in both structure and behavior. This division comprises orders such as:

- Neuroptera (Lacewings, Alderflies, snake flies)
- Mecoptera (Scorpion flies)
- Lepidoptera (Butterflies and moths)
- Diptera (Two-winged flies or true flies)
- and others.

Activity

1. Identify and collect insects belonging to Apterygota and Pterygota.
2. Observe insects from both Exopterygota and Endopterygota divisions to understand their unique characteristics.

3

Taxonomic Characteristics of the Order Orthoptera and Its Classification

Objective: The objective of this practical exercise is to understand the taxonomic features of the order Orthoptera and classify its suborders and families.

Observations: Order - Orthoptera (Orthos = straight, pteron = wing) Includes grasshoppers, crickets, and locusts.

Key Characteristics

1. Presence of wings, either fully developed, shortened (brachypterous), or absent (apterous).
2. Mouthparts are of the biting and chewing type (mandibulate).
3. Hind legs are typically enlarged and adapted for jumping.
4. Two pairs of wings, sometimes reduced or vestigial. The front wings, called Tegmina, are straight and thickened, while the hind wings are membranous.
5. Exhibits gradual metamorphosis, where nymphs closely resemble adults in essential features and habits.
6. Presence of a pair of unsegmented short cerci.

The order is further divided into two suborders

Suborder 1: Ensifera

- Antennae are longer than the body length and segmented.
- Tympanal organs (auditory organs) are located on the tibia of the legs.
- Examples include long-horned grasshoppers and crickets.

Suborder 2: Caelifera

- Antennae are shorter than the body length, typically with fewer than thirty segments.

- Tympanal organs are located at the sides of the 1st abdominal segment.
- Examples include short-horned grasshoppers and locusts.

Family: Acrididae

1. Moderate-length insects with prominent head regions.
2. Active during the day (diurnal).
3. Antennae are always shorter than the body length.
4. Auditory organs are situated on the sides of the 1st abdominal segment.
5. Typically, there is one generation per year.

Examples include Kharif grasshopper (*Hieroglyphus banian, Hieroglyphus nigroreplatus*), Desert locust (*Schistocerca gregarea*), and Migratory locust (*Locusta migratoria*).

Activity

1. Create neatly labeled diagrams of representative insects from the Orthoptera order.
1. 2. Collect the listed insects and observe them to identify different key family characteristics.

4

Taxonomic Characteristics of the Order Isoptera and Its Classification

Objective: The aim of this practical session is to explore the taxonomic features of the order Isoptera and classify its families.

Order: Isoptera (Iso = equal, ptera = wing) – Termites

Taxonomic Characteristics

1. Moderate-sized, social insects with thin skin, comprising various castes such as winged kings and queens, wingless kings and queens, workers, and soldiers.
2. Exhibits simple metamorphosis.
3. Mouthparts are of the typical biting and chewing type.
4. Wings are equal in size, long, narrow, membranous, and somewhat opaque.
5. Workers and soldiers of both sexes are wingless and sterile forms.

Families

- Mastotermitidae (e.g., *Mastotermes* spp.)
- Kalotermitidae (e.g., *Kalotermes* spp.)
- Rhinotermitidae (e.g., *Rhinotermes* spp.)
- Hodotermitidae (e.g., *Hodotermes* spp.)

Family: Termitidae

Characteristics

1. Members are predominantly subterranean and construct termitariums.
2. Wings are slightly reticulate, with the wing membrane and margin being more or less hairy.
3. Pronotum of workers and soldiers is narrow.

4. The queen exhibits enormous proportions, with the increase in size affecting only the abdomen and not the head and thorax. This enlargement is termed Physogastry.
5. Example: Termite - ***Odontotermes obesus / Microtermes obesi***

Activity

1. Collect and identify species of isopteran insects.
2. Create neatly labeled diagrams illustrating the different castes of termites.

5

Taxonomic Characteristics of the Order Thysanoptera and the Family Thripidae

Objective: To explore the habits, habitats, and taxonomic features of the order Thysanoptera and the family Thripidae.

Order: Thysanoptera

Etymology: Thysano (fringed) + Ptera (wings) = Fringed-winged Insects (THRIPS)

Habitat: Terrestrial, commonly found in flowers

Habit: Phytophagous, feeding on plant material

Size: Minute to small, characterized by thin, slender elongated bodies

Taxonomic Characteristics

1. **Antennae:** Typically 6-10 segmented, with sensorial structures (cone-like) on the 3rd or 4th segment.
2. **Compound Eyes:** Conspicuous, with 3-4 ommatidia in apterous forms and up to 150 in winged forms. Winged forms also exhibit 3 ocelli.
3. **Mouthparts:** Asymmetrical, lacking the right mandible. Adapted for lacerating, rasping, and sucking, with three stylets. The mouth cone, formed by the labrum, labium, and maxillae, extends ventrally between the fore coxae.
4. **Wings:** Thrips may be winged or wingless. Flight-capable thrips possess two pairs of similar, strap-like wings with a ciliated fringe on the margins, from which the order derives its name. Fully developed wings are long and narrow with highly reduced venation (often with 1-2 veins).
5. **Legs**: Typically terminating in two tarsal segments, with a bladder-like structure known as an arolium at the pretarsus.

6. **Abdomen:** Elongate, comprising 10-11 segments, usually tapering posteriorly.
7. **Anal Cerci:** Absent

Metamorphosis: Intermediate, exhibiting characteristics between simple and complex. The 1st and 2nd instar larvae are active and resemble adults (referred to as nymphs), while the 3rd and 4th instars consist of inactive pre-pupal and pupal-like stages.

Parthenogenesis: Common, with males rarely observed.

Family: Thripidae

Species

- **Chilli Thrips:** *Scirtothrips dorsalis*
- **Onion Thrips:** Thrips tabaci
- **Grapevine Thrips:** *Rhapiphorothrips cruentatus*

Key Features

1. Thripidae is the largest, most significant, and most injurious family within Thysanoptera.
2. Antennae typically comprise 6-9 segments, with the 3rd and 4th segments being conical and bearing sensoria. The antennae often feature an apical style with 1-2 segmented apical styles, and the 4th segment is usually enlarged.
3. Members of this family may exist in winged or wingless forms. If winged, their wings are narrow, pointed at the tip, and fringed with hairs along the margins.
4. Thripidae can be distinguished from other thrips by the presence of a saw-like ovipositor curving downwards.
5. The first and second segments of the tarsi exhibit claw-like appendages.

Activity

1. Collect the listed insects and observe them to identify various taxonomic characteristics.
2. Create a neatly labeled diagram illustrating the morphology of thrips.

6

Taxonomic Characteristics of Order Neuroptera and Its Family Chrysopidae

Objective: To identify key characters of the order Neuroptera and explore the characteristics of the family Chrysopidae.

Order: Neuroptera

Etymology: Neuro (nerve) + Ptera (wings) = Wings with net-like patterns of veins

Common Members: Lacewings, Aphid Lions, Ant Lions, Alder Flies, Snakeflies

Key Features of Neuroptera

1. Neuroptera possess two pairs of similar-sized membranous wings with a complex, net-like pattern of veins.
2. They are relatively fragile insects and weak fliers.
3. Antennae are filiform, with or without a terminal club.
4. Both adults and larvae have chewing mouthparts, with larval mandibles being very strong and elongated. Some larval mouthparts are modified for piercing and sucking.
5. Larvae are compodiform and predaceous.
6. Lacewings and their immature forms, known as aphid lions, are prevalent in this order. Both adults and larvae feed on aphids. Immature ant lions, also known as "doodlebugs," create pits in dry, dusty soil.
7. Adult green lacewings can be found throughout the year and are considered beneficial because they feed on other insects.
8. Six out of eight malpighian tubules are modified as silk glands.
9. Larvae spin cocoons through the anal spinneret, and pupation takes place in a silken cocoon.

Family: Chrysopidae (Green Lacewings, Aphid Lions, Golden Eyes)

Key Characteristics

1. Body color is pale green.
2. Eggs are mounted on stalks to avoid predation and cannibalism.
3. Larvae prey on soft-bodied insects, especially aphids, often carrying a layer of debris on their bodies for camouflage.
4. They emit a stinking fluid when alarmed, originating from prothoracic stink glands.
5. These insects are mass multiplied and released in fields for pest control.

Activity

1. Draw neat labeled diagrams of a green lacewing.
2. Collect the listed insects and observe them to identify different key characters.

7

Taxonomic Characteristics and Classification of Order Coleoptera and Agriculturally Important Families

Objective: To understand the taxonomic characters and classification of the order Coleoptera, which includes beetles and weevils.

Order: Coleoptera (Coleos = sheath, Pteron = wing) Beetles and Weevils

Taxonomic Characteristics

1. Two pairs of wings, with the forewings thickened and hardened, forming protective covers called elytra, while the hind wings are membranous and protected by the elytra.
2. Both larvae and adults possess biting and chewing mouthparts.
3. Metamorphosis is complete.
4. Beetle larvae are commonly known as grubs. Snout beetle (weevil) grubs are legless (apodous).

This order is divided into the following two suborders

Suborder 1: Adephaga

Key Characters

1. Beetles in this suborder are mostly predatory, feeding on other insects.
2. Antennae are generally filiform.
3. Notopleural suture is present.
4. The first visible abdominal sternum is divided by the hind coxae, and the posterior margin of this sternum does not extend completely across the abdomen.

Important Families

1. **Family: Cicindellidae**

Example: Tiger beetle (*Cicindella sexpunctata*)

2. **Family: Carabidae**

Examples

- Carabid beetle (*Anthia sexguttata*)
- Carabid beetle (*Chlaenius bioculatus*)
- Carabid beetle (*Calosoma indica*)

Suborder 2: Polyphaga

Key Characteristics

1. The first visible abdominal sternum is not divided by the hind coxae, and the posterior margin of this sternum extends completely across the abdomen.
2. Hind trochanters are small.
3. Notopleural suture is absent.

Important Families

1. **Family: Dermestidae**

   Example: Khapra beetle (*Trogoderma granarium*)

2. **Family: Curculionidae** (curculioni = weevils or snout beetles)

Examples

- Rice weevil (*Sitophilus oryzae*)
- Gujhia weevil (*Tanymecus indicus*)
- Sweet potato weevil (*Cylas formicarius*)

3. Family: Bruchidae

   Example: Pulse beetle (*Callosobruchus chinensis*)

4. Family: Chrysomelidae

Examples

- Red pumpkin beetle (*Raphidopalpa foveicollis*)
- Rice hispa (*Decladispa armigera*)

- Singhara beetle (*Galerucella bermanica*)

5. Family: Tenebrionidae

   Example: Rust red flour beetle (*Triboliuam castaneum*)

6. Family: Coccinellidae

   Subfamily: Coccinellinae (lady bird beetles)

   Example: Lady bird beetle (*Coccinella septempunctata*)

7. Family: Melolonthidae (Scarabaeidae)

   Example: White grub (*Holotrichia consanguinea*)

Activity

1. Draw neat labeled diagrams of different insects under Suborder Adephaga and Polyphaga.
2. Collect and identify the beetles and weevils.

8

Taxonomic Characteristics of Order Hymenoptera and Agriculturally Important Families

Objective: To comprehend the fundamental taxonomic characteristics of the order Hymenoptera and its agriculturally significant families.

The order Hymenoptera displays remarkable diversity in its insect inhabitants. The term "**Hymenoptera**" originates from "**Hymen**" meaning membrane or the God of marriage, and "**ptera**" meaning wings.

Taxonomic Characters

1. Hymenopterans possess mandibulate mouthparts, where the labium and maxilla's basal sections lie closely side by side, forming a sharp transverse fold near the basal third. In bees, these structures integrate to form a lapping tongue, representing a chewing and lapping type of mouthpart.
2. The thorax is specialized for efficient flight, featuring a collar-like pronotum, an enlarged mesothorax, and a small metathorax. The prothorax and metathorax are fused with the mesothorax.

The order presents the following key characteristics

1. Two pairs of membranous wings, with forewings larger than hind wings, free from scales or dense hairs.
2. Simplified wing venation, with forewings and hind wings coupled by hooklets (hamuli) along the leading edge of the hind wing.
3. A small roundish sclerite, the tegula, covers the base of the forewing.
4. The abdomen is basally constricted, with the first abdominal segment fused with the metathorax. The second segment, known as the pedicel, connects the thorax to the abdomen. The portion beyond the pedicel is termed the gaster or metasoma.

5. Females possess a true primitive ovipositor with three pairs of valves, variously modified for oviposition, stinging, or piercing plant tissues.
6. Complete metamorphosis is observed, with larvae exhibiting varying characteristics.
7. Sex determination occurs through the fertilization of eggs, with fertilized eggs developing into females and unfertilized eggs producing males. Males are haploid, while females are diploid.

Agriculturally Important Families

1. **Ichneumonidae:** Known for their characteristic forewing venation and often elongated ovipositors.
2. **Braconidae:** Resembling ichneumonids but differing in wing venation and abdominal structure.
3. **Trichogrammatidae:** Identified by their distinctive tarsi and antennal characteristics.
4. **Eulophidae:** Recognizable by their metallic coloration and segmented tarsi.
5. **Platygasteridae:** Minute black insects with distinctive antennal structure.
6. **Apidae:** Notable for honey bees, with specialized mouthparts and pollen-collecting legs.
7. **Tenthredinidae:** Characterized by their saw-toothed ovipositors and eruciform larvae.

Activity

1. Collect the listed insects and identify them based on their distinct family characteristics.
2. Collect and identify bee species, providing well-labeled diagrams.

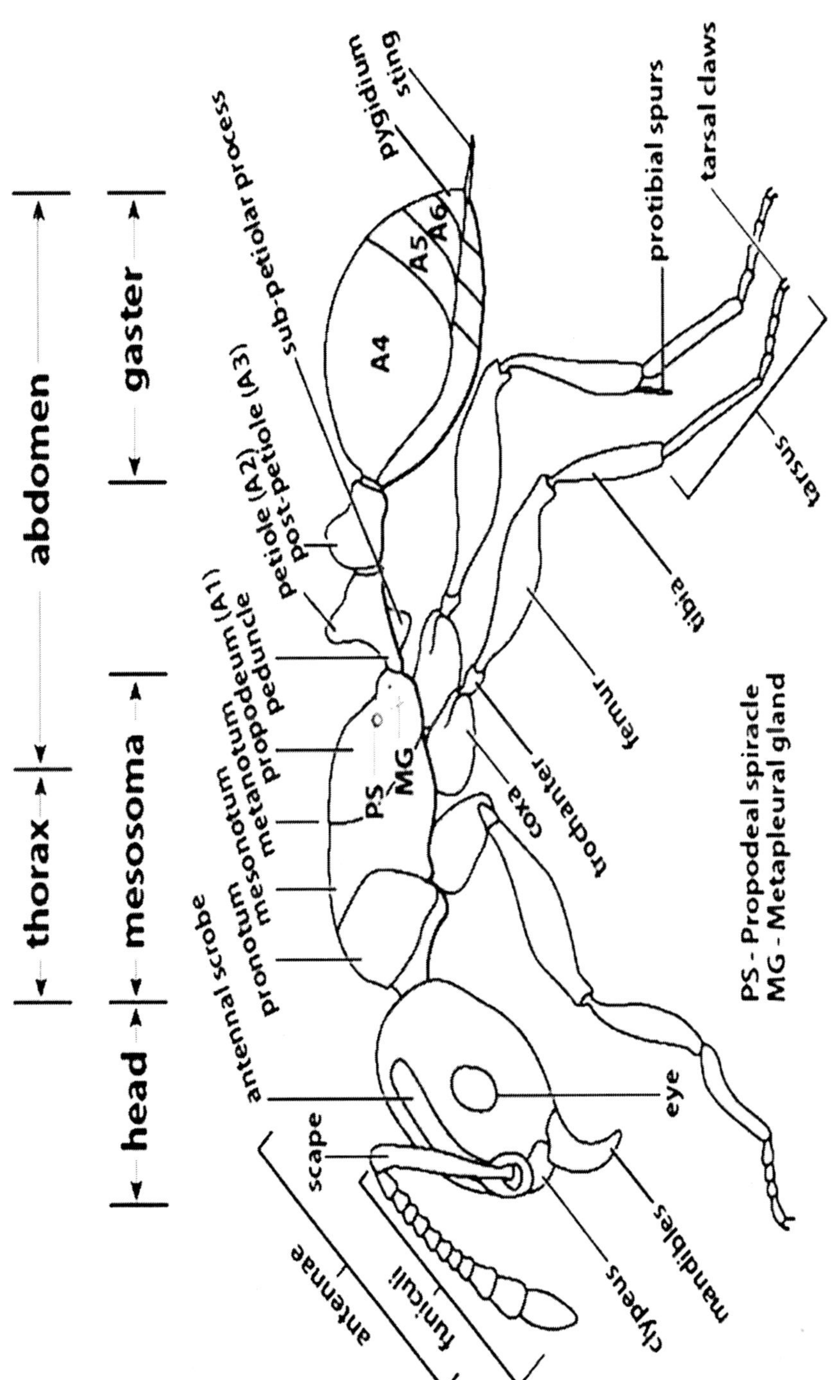
head
thorax
abdomen
mesosoma
gaster
antennal scrobe
pronotum
mesonotum
metanotum
propodeum (A1)
peduncle
petiole (A2)
post-petiole (A3)
sub-petiolar process
pygidium
sting
A4
A5
A6
PS
MG
protibial spurs
tarsal claws
tarsus
tibia
femur
trochanter
coxa
eye
scape
antennae
funiculi
clypeus
mandibles
PS - Propodeal spiracle
MG - Metapleural gland

9

Comparative Characteristics of Symphyta and Apocrita

Objective: Explore and understand the key identical characteristics that differentiate the suborders Symphyta and Apocrita within the order Hymenoptera.

Introduction: Hymenoptera, a large and diverse order of insects, is divided into two main suborders: Symphyta and Apocrita. These suborders exhibit distinct physical and behavioral traits. Understanding these differences is crucial for entomological studies, particularly in areas like pest control, ecology, and conservation.

Comparative Characteristics

Characteristic	Symphyta	Apocrita
Abdominal Attachment	Broadly joined to the thorax	Petiolated (narrow waist)
Larval Form	Caterpillar-like, eruciform type	Grub-like, apodous eucephalous type
Presence of Stemmata	Present	Absent
Legs in Larvae	Both thoracic and abdominal legs present	Legs absent
Ovipositor Type	Saw-like, suited for piercing plants	Modified, not saw-like, often stinging
Behavioral Complexity	Less sophisticated	More sophisticated
Feeding Habits	Phytophagous (plant-eating)	Generally parasitic
Examples	*Athalia lugens proxima*	*Campoletis chlorideae, Bracon brevicornis, Trichogramma spp.*

Activity

1. Drawing and Labeling Diagrams:
    - Create detailed, labeled diagrams of representative insects from the suborders Symphyta and Apocrita.
    - Highlight the distinguishing features mentioned in the table to help in visual differentiation.

10

Taxonomic Characteristics of the Order Lepidoptera and Agriculturally Important Families

Objective: Identify and understand the distinctive features of the order Lepidoptera and its significance in agriculture through key families.

Introduction: Lepidoptera, comprising butterflies and moths, is characterized by scaly wings. The order's name, derived from the Greek words 'lepido' for scales and 'ptera' for wings, reflects this defining trait. These insects are crucial in ecological networks, serving roles in pollination and as a part of the food web, while some species are significant agricultural pests.

Key Characteristics

1. Morphology

- The bodies, wings, and appendages are densely covered with overlapping scales, providing color, rigidity, and strength.
- These scales also help insulate the body and optimize airflow during flight.
- Adult mouthparts are adapted for siphoning, featuring a coiled proboscis resembling a watch spring, used for feeding on nectar and stored beneath the head when not in use. Mandibles are typically absent.

2. Wing Structure

- Wings are membranous and covered with pigmented scales.
- The forewings are generally larger than the hind wings, with few cross veins.
- Wings are coupled via frenate or amplexiform mechanisms, aiding in flight stability.

3. Larval and Developmental Features

- Larvae are typically eruciform (caterpillar-like) with strong mandibles for chewing.

- They possess a set of lateral ocelli, short three-segmented antennae, and three pairs of thoracic legs ending in claws.
- Additionally, they have two to five pairs of fleshy, unsegmented prolegs with crochets at their ends, aiding in locomotion.
- Pupation occurs in an obtect state, where the pupa is usually encased in a cocoon composed of various materials like soil, frass, silk, or larval hairs.

Agriculturally Important Families

1. **Nymphalidae:** Brush-footed or four-footed butterflies
2. **Lycaenidae:** Blues, Coppers, Hair streaks
3. **Papilionidae:** Swallowtails
4. **Pieridae:** Whites and Sulphurs
5. **Satyridae:** Browns, Meadow-browns
6. **Arctiidae:** Tiger moths
7. **Bombycidae:** Silk worm moths
8. **Cochlididae:** Slug caterpillars
9. **Crambidae:** Grass moths
10. **Gelechiidae:** Paddy moths
11. **Geometridae:** Loopers
12. **Lymantridae:**Tussock moths
13. **Noctuidae:** Noctua moths
14. **Pyraustidae:** Grass borers
15. **Saturniidae:** Giant silk worm moths, Moon moths
16. **Sphingidae:** Hawk moths, Sphinx moths, Horn worms
17. **Hesperiidae:** Skippers

Activities

1. Collect and identify specimens from the butterfly and moth families listed, noting key distinguishing features.
2. Create detailed diagrams to highlight the morphological differences between the families, focusing on wing patterns, scale arrangement, and larval forms.

11

Taxonomic Characteristics of Order Diptera and Agriculturally Important Families

Objective: This practical aims to familiarize students with the essential traits of Diptera and explore the agriculturally significant families within this order.

Introduction: Diptera, commonly known as true flies, encompass a diverse group of insects like houseflies and mosquitoes. Their unique characteristics and roles in ecosystems make them crucial to understand, especially in agricultural contexts where they can impact crop productivity.

Key Characteristics of Order Diptera

1. **Single Pair of Wings:** Diptera possess a distinctive feature of having only one pair of wings, with the hind pair transformed into halteres for balance.
2. **Mouthparts:** Dipterans exhibit diverse mouthpart adaptations, either suited for piercing and sucking or for sponging, depending on their feeding habits.
3. **Thorax Structure:** Their thorax structure is notable for the fusion of the prothorax and metathorax, supporting a well-developed mesothorax responsible for wing and haltere attachment.
4. **Metamorphosis:** Dipterans undergo complete metamorphosis, progressing through egg, larval (often apodous or legless), pupal, and adult stages.
5. **Larvae:** Dipteran larvae, commonly referred to as maggots, typically possess biting and chewing mouthparts, adapting to various ecological niches.
6. **Pupal Stage:** The pupal stage may either be free or enclosed within a puparium, facilitating the transformation from larva to adult.

Agriculturally Important Families of Diptera

1. Culicidae (Mosquitoes)
    - Examples include *Anopheles spp*. and *Culex fatigans*.
2. Cecidomyiidae (Gall Midges)
    - Noteworthy species encompass the rice gall midge (*Orseolia oryzae*), mango gall midge (Dasyneura mangiferae), and linseed gall midge (*Dasyneura lini*).
4. Tabanidae (Horse Flies)
    - The horse fly (*Tabanus maculicornis*) stands as a representative of this family.
5. Asilidae (Robber Flies)
    - Robber flies, such as Philonicus albiceps, are prominent members of this family.
6. Bombyliidae (Bee Flies)
    - The bee fly (*Bombylius major*) is an example of this family.
7. Tephritidae (Fruit Flies)
    - Species like the melon fruit fly (*Bactrocera cucurbitae*) belong to this family.

Activities

1. Drawing Labeled Diagrams:
2. Insect Collection and Observation

12

Comparative Taxonomic Characteristics of Heteroptera and Homoptera

Objective: To investigate and compare the taxonomic traits of Heteroptera and Homoptera, two suborders within the order Hemiptera, commonly known as true bugs.

Introduction: Heteroptera and Homoptera are two important suborders within the order Hemiptera, commonly known as true bugs. Understanding their taxonomic characters is essential for classification and identification purposes.

Taxonomic Characteristics

Character	Heteroptera	Homoptera
Examples	True Bugs	Jassids, Leafhoppers, Aphids, Whiteflies, Scales, Mealybugs
Basic Character	Hetero: Different wings	Homo: Same/uniform/alike wings
Size	Comparatively large	Comparatively small
Head	Prognathous	Hypognathous / Opisthognathous
Antenna	4-5 segmented	3-10 segmented
Rostrum	Arises from the front part of the head	Arises from the posterior part of the head; head base extends between anterior coxa
Pronotum	Large pronotum, relatively small meso- and metanotum	Small pro-notum, collar-like; large meso-notum, small meta-notum
Forewings	Hemi-elytra; hard forewings	Uniform in consistency; entire forewings
Wings Position	Held flat over the body at rest	Held roof-like/tent-like over the body at rest
Tarsi	3 segmented	1-3 segmented
Sound Production	No sound-producing organs	Many can produce sounds; sound production due to tymbal organs in the first two abdominal segments (Cicadas)
Glands	Odoriferous glands present; opens at hind coxae, giving off characteristic odor	Wax glands present on the entire body

Character	Heteroptera	Homoptera
Life History	Parthenogenesis present in some insects	Complex life history involving sexual and parthenogenetic generations, winged and wingless individuals
Important Families	Cimicidae (Bed bugs), Pentatomidae (Stink/Shield Bugs), Lygaeidae (Seed bugs, Chinch Bugs), Miridae (Mirid bugs)	Cicadellidae (Leafhoppers, Jassids), Delphacidae (Planthoppers), Aphididae (Aphids or Plant Lice), Pseudococcidae (Mealybugs), Pyrrhocoreidae (Cotton bugs/stainers), Coridae (Leaf-footed bugs)
		Coccidae (Scale insects), Aleurodidae (Whiteflies), Lophodidae (Aeroplane bugs)

Activity

1. Drawing Labeled Diagrams:

   - Create detailed, labeled diagrams of different insects belonging to Heteroptera and Homoptera, highlighting their distinguishing features.

2. Identification of Economically Important Homopteran Family:

Identify and discuss the economically significant family within the Homoptera suborder, focusing on its impact on agriculture or other relevant industries.

13

Comparative Taxonomic Characteristics of Beetles and Weevils

Objective: To explore and compare the taxonomic traits of beetles and weevils through detailed examination and suitable examples.

Introduction: Beetles and weevils are fascinating insects with distinctive characteristics that set them apart. This practical aims to explore and understand the taxonomic traits of both beetles and weevils through comparison

Taxonomic Characteristics of Beetles and Weevil with suitable example

Sl. No.	Beetle	Weevil
1.	Mouth parts typically chewing type	Mouth parts chewing type but modified into snout-like structure
2.	Both pairs of wings present; fore wing elytrate, resting in straight suture when at rest	Hind wing absent; fore wing elytrate, immovably united
3.	Antennae capitate/serrate/lamellate	Antennae clavate
4.	Tarsi 3 to 5 segmented	Tarsi 4 segmented

Activity

1. Collecting and Identifying Beetles and Weevils

14

Comparative Taxonomic Characteristics of Butterflies and Moths

Objective: Aims to investigate and understand the distinguishing features of butterflies and moths, focusing on their unique traits and examples.

Introduction: Butterflies and moths are fascinating insects belonging to the order Lepidoptera. By examining their key characteristics, we can distinguish between these two groups and appreciate their ecological roles.

Character	**Butterflies**	**Moths**
Shape and Structure of Antenna	Thin filamentous antennae, club-shaped at the end	Comb-like feathery antennae, filamentous, or club-shaped
Wing Coupling Mechanism	Amplexiform type of wing coupling mechanism	Frenulum filamentarising from hind wing coupling with barbs on forewing; some have a lobe on forewing called jugum
Pupae	Form exposed pupae (chrysalis)	Pupa enclosed in cocoon or shell
Coloration of Wings	Brightly colored	Usually brown, gray, white, or black with obscuring patterns for camouflage
Body Structure	Fine scales, slender, smoother abdomens	Stout, hairy or furry-looking bodies, dense and fluffy scales on wings
Behavioral Difference	Diurnal	Nocturnal
Resting Posture	Fold wings above abdomen when perched	Rest with wings spread out to sides
Economic Importance	Pollinators in adult stage, weed killers in larval stage	Larval stages feed on cultivated crops, becoming important pests
Example	Blue butterfly	Fruit-sucking moth

Activity

1. Drawing Labeled Diagrams of Butterflies and Moths

15

Comparative Taxonomic Characteristics of Megaloptera and Planipennia

Objective: This practical aims to understand the taxonomic traits of Megaloptera and Planipennia, focusing on their unique features and differences.

Introduction: Megaloptera and Planipennia are two distinct orders of insects with distinct characteristics. By examining their taxonomic traits, we can better understand their classification and ecological significance.

Character	Megaloptera	Planipennia
1. Body size	Large	Small
2. Antennae	Large, filiform, short	Small; very long, many-segmented, filiform, moniliform, pectinate
3. Wings	Without pterostigma	With pterostigma;
4. Hind wings	Long but not exerted	without large and fold.
5. Veins	Unbranched at margin; branches or veins bifurcated	net-like venation branched at margin
6. Ovipositor	Long but not exertet	Plough- like
7. Larval mouth parts	Biting mouthparts; sickle-like mandibles	Piercing, suctorial mouthparts
8. Larval habitat	Aquatic	Rarely aquatic (Sisyridae)

Activity

1. Drawing Labeled Diagrams of Megaloptera and Planipennia:

    - Students will create neat and labeled diagrams of Megaloptera and Planipennia, highlighting the key identical characteristics mentioned above.

16

Taxonomic Characteristics of Terebrantia and Tubilifera and Agriculturally Important Families

Objective: This practical aims to understand the taxonomic traits of Terebrantia and Tubilifera, focusing on their unique features and differences.

Introduction: Terebrantia and Tubilifera are two suborders of thrips, tiny insects belonging to the order Thysanoptera. By examining their taxonomic traits, we can better understand their classification and ecological significance.

Taxonomic Characteristics

Suborder Terebrantia

1. Female bearing saw-like serrated ovipositor.
2. Female with conical abdominal apex, while male with blunt and round abdominal apex.
3. Forewings large with at least one complete longitudinal vein.

Family Thripidae (Terebrantia)

- Ovipositor curved downwards.
- Wings narrow and pointed; antennae 6-9 segmented, with the 6th segment being the largest.
- Last abdominal segment of female is conical.

Suborder Tubilifera

1. Female without ovipositor.
2. 10th abdominal segment is tubular.
3. Fore and hind wings are similar, with no veins or with one vestigial vein.
4. No longitudinal veins run the full length of a wing. Wings without hairs (microtrichia).

Family Phlaeothripidae (Tubilifera)

- Antennae with 7-8 segments, with the 3rd segment being the largest.
- Head rounded; maxillary palp 2-segmented.

Activity

1. Collecting and Identifying Terebrantia and Tubilifera Insects:

17

Taxonomic Characteristics of Nematocera and Brachycera Insects

Objective: Aims to understand and identify the key differences between insects belonging to the suborders Nematocera and Brachycera.

Introduction: Nematocera and Brachycera, both part of the Diptera order, represent flies with unique body shapes and antennal lengths: slender and long for Nematocera, and stout with shorter antennae for Brachycera. These distinctions are crucial for understanding their ecological roles across diverse ecosystems and their impact on agriculture and public health.

1. Drawing Labeled Diagrams

    - Create neat and labeled diagrams of representative insects from the suborders Nematocera and Brachycera. Highlight the key characteristics mentioned below.

Key Characteristics

Characters	Nematocera	Brachycera
I. **Larva**		
1. Head	1. Eucephalous,fully developed except in cecidomyidae and Tipulidaess	1. Hemicephalous, reduced represented by only lowcapsule
2. Antennae	2. 1-6 segment well- developed	2. Reduced, papillae- like welldeveloped, mandible sarticulate vertically.
3. Mouth parts	3. Well developed ,mandible articulate transversely.	3. Slightly convolulated
4. Alimentary Canal	4. Short – straight	4. usually more long and coiled
5. Instars	5. 4,(exception-6 in simulidae)	5. Generally undergo 3 instars during their development
II. Pupa		
6. Type	6. Obtect	6 Usually obtect or adectous
III. Adult		
7. Antennae	7. Typically longer, filiform (thread-like), with multiple segments	7. Usually shorter, with fewer segments, often aristate (bristle-like)

Characters	Nematocera	Brachycera
8. Palpi	8. 4 or 5 segment, pendulous	1 or 2- segmented porrect twice bent (exc.- Acroceridae)
9. Pleural suture of mesothorax	9. 4 or 5 segemented, pendulous straight, exception- psychodidae	9. Present
10.Wing discal ccell	10. Absent	10. Closed
11. Cubital cell	11.Widely open	11. blade like piercing
12. Mouthparts	12.Stylet- like, piercing type, only in female	12. Type only in female
13. Salivary glands	13.Tubular, trilobed, thoracic	13.Tubular running upto some length of abdomen

Activity

1. Draw neat labeled diagrams of representative insects from the insect suborder viz., Nematocera

18

Taxonomic Characteristics of Order Mantophasmatodea and its Families

Objective: This practical aims to understand the distinctive features of the insect order Mantophasmatodea and its major family, Mantophasmatidae.

Introduction: Mantophasmatodea is a small insect order, characterized by its wingless nature and uniform body shape. Understanding its key characteristics helps in the identification and classification of these fascinating insects.

Key Characteristics

1. Mantophasmatodea is the smallest insect order, resembling certain grasshoppers or stick insects.
2. The body length ranges from 9 to 24 mm, with males usually smaller than females.
3. Body coloration varies, typically in shades of brown, gray, green, or yellow, with mottled patterns and longitudinal stripes.
4. Nymphs resemble adults.
5. Well-developed compound eyes, but lacking ocelli.
6. Chewing mouthparts directed downward.
7. Abdomen consists of 10 well-developed segments and a reduced eleventh segment.

Major Family

Mantophasmatidae is the sole family within the order, containing three genera.

Activity

1. Observation of Taxonomic Characters:
    - Collect specimens of Mantophasmatodea insects and observe them for their taxonomic characteristics. Note the body size, coloration patterns, presence of compound eyes, mouthparts, and abdominal structure.

- Through this activity, you will gain practical experience in identifying and understanding the key characteristics of Mantophasmatodea insects and their major family, Mantophasmatidae.

19

Identifying Locust and Grasshopper Based on Morphometric and Behavioral Characteristics

Objective: To differentiate between locusts and grasshoppers based on their morphometric measurements and behavioral traits.

Introduction: Locusts and grasshoppers are both members of the order Orthoptera, and while they share many similarities, they also exhibit distinct differences, particularly in behavior and coloration. Understanding these differences is crucial for accurate identification and management, especially considering the significant economic impact locust swarms can have on agriculture.

Locust

2. Description: Locusts are short-horned grasshoppers belonging to the family Acrididae. They undergo a behavioral and physiological change known as phase polyphenism, transitioning from solitary to gregarious behavior in response to population density.
3. Behavior: Solitary locusts exhibit typical grasshopper behavior, while gregarious locusts form swarms and display synchronized movements.

Fig. Locust

4. Coloration: Gregarious locusts often develop brighter colors, and they can change their coloration, form, physiology, behavior, and fertility during swarming phases.

5. Morphometrics: Morphometric differences may include variations in body size, wing length, and leg structure between solitary and gregarious phases.

Grasshopper

1. Description: Grasshoppers are long-horned insects known for their plant-eating habits. They have elongated hind legs adapted for jumping.
2. Behavior: Grasshoppers are typically solitary insects, although they may aggregate in certain situations. Their behavior is generally consistent, with individuals feeding and moving independently.
3. Coloration: Grasshoppers typically have brown, grey, or green coloration, providing camouflage to blend with their environment. Males may exhibit brighter colors to attract females.
4. Morphometrics: Grasshoppers generally have medium-sized bodies ranging from 1 to 7 cm in length. They possess characteristic long hind legs for jumping and a distinct body shape.

Fig. Grasshoper

Similarities

1. Both locusts and grasshoppers are herbivorous insects.
2. They belong to the order Orthoptera and share common features such as large compound eyes, short antennae, and strong mandibles.
3. Both insects undergo incomplete metamorphosis, progressing through egg, nymph, and adult stages.
4. They have elongated hind legs adapted for jumping and possess two pairs of wings, with two membranous wings in the back.

Activity

1. Collecting and Identifying Morphometric and Behavioral Characteristics:
    - Gather specimens of locusts and grasshoppers and observe their morphometric features, such as body size, wing length, and leg structure. Additionally, note their behavior, including feeding habits, movement patterns, and responses to stimuli.

20

Taxonomic Characteristics of Order Dictyoptera

Objective: The order Dictyoptera includes insects like cockroaches and praying mantises with netted wings. Recognizing their unique taxonomic traits is essential for precise identification and classification.

Background: The order Dictyoptera consists of insects with netted wings, including cockroaches and praying mantises. Understanding their taxonomic characteristics is essential for accurate identification and classification.

Taxonomic Characters

1. Head: Typically hypognathous (head directed downwards).
2. Antennae: Long and may be longer than the body, filiform in mantises and setaceous in cockroaches.
3. Mouthparts: Mandibulate, adapted for chewing and biting.
4. Wings:Two pairs of wings, with forewings modified into leathery tegmina and hind wings membranous with a large anal lobe folded fan-like.
5. Legs: Cockroaches have cursorial legs for running, while mantises have raptorial forelegs for grasping prey and normal middle and hind legs for running. Forelegs of mantises have a long coxa and spiny femur.
6. Anal Cerci: Present in both cockroaches and mantises, with many segmented cerci that are sensitive to air movements.
7. Eggs: Laid in ootheca, which is a protective capsule formed from foam secreted by female accessory glands.

Activity

1. Drawing Labeled Diagrams: Create neat and labeled diagrams of representative insects from the order Dictyoptera, highlighting their key taxonomic features.

2. Collecting and Identifying Dictyopteran Insects: Gather specimens of cockroaches and praying mantises and observe their morphological characteristics to identify them accurately. Pay attention to features such as body shape, antennae, wings, legs, and anal cerci.

21

Taxonomic Characteristics of Family Mantidae

Objective: To investigate and comprehend the taxonomic traits of family Mantidae, specifically cockroaches and praying mantises, through detailed examination and diagrammatic representation.

Introduction: The family Mantidae comprises a diverse group of insects known as mantises, known for their predatory behavior and distinctive morphology. Understanding their taxonomic characteristics is crucial for identification and classification.

Taxonomic Characters of Family Mantidae

1. Habitat: Mostly found in tropical and subtropical regions.
2. Diet: Predatory in nature, both nymphs and adults feed on other insects and small animals.
3. Cryptic Coloration: Often exhibit cryptic colorations, blending in with their surroundings for camouflage.
4. Body Shape: Long body with a flattened abdomen.
5. Head: Triangular in shape and often deflected downwards.
6. Ocelli: Three simple eyes present.
7. Antennae: Filiform, long, and composed of many segments.
8. Thoracic Structure: Prothorax greatly elongated, while meso and meta-thoracic segments are short.
9. Front Legs: Characterized by raptorial front legs with a long coxa, tibia, and femur bearing prominent spines.
10. Anal Cerci: Short and segmented.
11. Stridulatory Organs: No specialized stridulatory organs, though some species may have a single ear on the metathorax.

12. Egg Laying: Eggs are laid in a water-tight egg case called an ootheca, which is affixed to plants. The ootheca is formed from a frothy gum secreted by female glands and hardens upon exposure to air.
13. Nymphs: Nymphs emerging from the ootheca often resemble ants.
14. Examples: Common mantis species include the Greenhouse Mantis (*Mantis religiosa*) and the Common Indian Mantis (*Gongylus gongyloides*).

Activity

1. Collecting and Identifying Taxonomic Characters: Gather specimens of mantid insects and observe their morphological characteristics closely. Note features such as body shape, coloration, head structure, antennae, leg morphology, and presence of ootheca. Document these features for identification and classification purposes.

22

Taxonomic Characteristics of Family Blattidae

Objective: To explore and document the taxonomic characteristics of cockroaches belonging to the family Blattidae through detailed observation and analysis

Introdcution: The family Blattidae comprises a group of nocturnal and omnivorous insects commonly known as cockroaches. Understanding their taxonomic characteristics is essential for identification and management in various environments.

Taxonomic Characteristics of Family Blattidae

1. Nocturnal and Omnivorous: Cockroaches are nocturnal insects that feed on a wide variety of organic matter, including food, clothing, and paper. They often impart a foul smell to food by contaminating it with their excreta.
2. Body Shape: Dorsoventrally flattened, depressed, and oval-shaped.
3. Pigmentation: Body heavily pigmented.
4. Head Structure: The head is not mobile in all directions and is often hidden by the pronotum.
5. Pronotum: Large plate-like or shield-like structure covering the head.
6. Ocelli: Ocelli are degenerated and sensitive to light, referred to as fenestrae.
7. Antennae: Setaceous or whip-like, long, and composed of many segments.
8. Legs: All three pairs of legs are similar and adapted for cursorial (running) locomotion.
9. Wings: Cockroaches have two pairs of functional wings used for flying. The forewings are modified as tegmina, while the hind wings are membranous with a large anal area.

10. Eggs: Eggs are arranged in a double row and enclosed in a sac called an ootheca, formed by secretions (polystyrene) from collateral glands, which are modified accessory glands of females.

11. Example: An example of a cockroach belonging to the family Blattidae is the American cockroach (Periplaneta americana).

Activity

1. Collecting and Identifying Taxonomic Characters: Gather specimens of cockroaches and carefully observe their morphological features. Note characteristics such as body shape, pigmentation, head structure, antennae morphology, leg structure, wing characteristics, and the presence of ootheca. Document these features for accurate identification and classification within the family Blattidae.

23

Taxonomic Characteristics of Family Pentatomidae

Objective: To explore and record the taxonomic features of insects in the family Pentatomidae through thorough examination and analysis of their morphology.

Introduction: The family Pentatomidae consists of medium to large-sized insects commonly known as stink or shield bugs. Understanding their taxonomic characteristics is essential for accurate identification and management in various environments.

Taxonomic Characteristics of Family Pentatomidae

1. Size and Shape:Medium to large-sized insects with broad shield-like bodies.
2. Body Structure: Medium-sized, shield-like, and brightly colored.
3. Ocelli: Ocelli are present, contributing to their sensory capabilities.
4. Head Structure: The lateral margins of the head conceal the bases of the antennae.
5. Antennae: Five-segmented antennae.
6. Pronotum: Broad and shield-shaped, with a large triangular scutellum on the mesonotum that sometimes extends posteriorly to cover the wings entirely.
7. Forewings:Known as hemi-elytra, with a large membranous corium and many longitudinal veins arising from a vein parallel to the apical margin of the corium.
8. Tarsi: Two to three segmented tarsi, with claws equipped with pulvilli.
9. Odor Production: Stink bugs produce highly disagreeable odors for defense. They have a pair of odoriferous glands opening at the hind coxae, and nymphs have four pairs of odoriferous glands on the dorsum of the abdomen.

10. Eggs:Typically barrel-shaped with spines on the upper end.

Examples

- Green Stink Bug: *Nezara viridula*
- Red Pumpkin Bug: *Aspongopus* (*Coridius*) *janus*
- Cabbage Painted Bug: *Bagrada cruciferarum*

Activity

1. Collecting and Identifying Taxonomic Characters: Gather specimens of pantid bugs and green stink bugs. Carefully observe their morphological features, including body shape, coloration, head structure, antennae morphology, wing

24

Taxonomic Characteristics of Family Cicadellidae

Objective: To thoroughly examine and record the fundamental taxonomic traits of insects belonging to the family Cicadellidae through detailed observation and analysis.

Background: The family Cicadellidae comprises economically significant insects known as leafhoppers. Understanding their key taxonomic characteristics is crucial due to their role as vectors of plant diseases and their impact on agriculture.

Key Characteristics of Family Cicadellidae

1. Economic Importance: Cicadellidae is economically significant as vectors of important plant diseases, causing plant injury by injecting toxins and resulting in hopper burns.
2. Size and Shape: Leafhoppers are typically small, slender insects with a wedge-shaped body, tapering posteriorly. They adopt a wedge-shaped posture during rest.
3. Jumping Behavior: Leafhoppers can jump several feet when disturbed. Both nymphs and adults exhibit a characteristic habit of running sideways or diagonally.
4. Antennae: Minute, bristle-like antennae with three segments.
5. Forewings: The forewings are somewhat thickened and often brightly colored.
6. Wing Veins: Forewing anal veins usually separate from the base to the anal margin, with anal veins 1A and 2A not uniting to form a Y-shaped vein.
7. Hind Tibia: Typically, one or two rows of small spines are present on the hind tibia.
8. Ovipositor: Well-developed ovipositor adapted for lacerating plant tissues for egg-laying.

9. Honeydew Production: Many leafhoppers excrete honeydew through their anus, leading to the development of sooty mold (black), which hampers the plant's photosynthesis ability.

Examples

- Cotton Leaf Hopper: *Amrasca biguttula biguttula*
- Paddy Leaf Hopper: *Nephotettix virescens*
- Mango Leaf Hopper: *Idioscopus clypealis, Amritodus atkinsoni*
- Bhendi Leaf Hopper: *Amrasca biguttula biguttula*

Activity

1. Collecting and Observing: Gather specimens of the listed leafhoppers and carefully observe them for different key family characters. Document their size, shape, behavior, antennae morphology, wing characteristics, presence of wing veins and spines, ovipositor structure, and honeydew production. This observation will aid in the accurate identification and classification of leafhoppers within the family Cicadellidae.

25

Taxonomic Characteristics of Family Aphididae

Objective: The objective of this practical exercise is to explore and document key taxonomic characteristics of Aphididae, focusing on specimens like cotton aphids (*Aphis gossypii*) and bean aphids (*Aphis craccivora*).

Introduction: The family Aphididae comprises economically significant insects known as aphids. Understanding their key taxonomic characteristics is crucial due to their impact on agriculture, acting as vectors of plant diseases and causing damage by feeding on various parts of plants.

Key Characteristics of Family Aphididae

1. **Economic Importance:** Aphididae is economically significant, being polyphagous and some acting as vectors of plant diseases while others feed on roots.
2. **Body Shape:** Aphids are pear-shaped, small, and soft-bodied insects typically found in large colonies, sucking sap from various parts of plants.
3. **Rostrum:** Aphids typically have a long and well-developed rostrum for piercing plant tissues and sucking sap.
4. **Antennae:** Fairly long antennae.
5. **Cornicles:** A pair of cornicles on the dorsal surface of the 5th and 6th abdominal segments. These cornicles produce a wax substance that protects the aphids from other insects.
6. **Alary Polymorphism:** Aphids exhibit alary polymorphism, with both winged and wingless forms. Wingless forms are predominant, but in winged forms, hind wings are much smaller with fewer veins.
7. **Wing Position:** At rest, aphids typically hold their wings vertically above their bodies, resembling a tent.
8. **Tarsus:** Tarsi are three-segmented with a pair of claws.

9. **Honeydew Production:** Aphids excrete honeydew through their anus, consisting of excess sap, sugars, and waste materials. Ants are attracted to honeydew, and the transportation of aphids by ants is called trophobiosis.
10. **Reproduction:** Reproduction in aphids can occur through parthenogenesis, oviparity, or viviparity. In aphid populations, the sexes are often unequally developed, with males being rare.

Examples

- Cotton Aphid: *Aphis gossypii*
- Bean Aphid: *Aphis craccivora*
- Apple Aphid: *Eriosoma lanigerum*

Activity

1. Collecting and Identifying: Gather specimens of cotton aphids and bean aphids. Identify and observe their taxonomic characters, including body shape, rostrum morphology, presence of cornicles, wing polymorphism, wing position at rest, tarsal structure, and honeydew excretion. Document these observations to facilitate accurate identification and classification of aphids within the family Aphididae.

26

Taxonomic Characteristics of Family Aleurodidae

Objective: To study and identify key taxonomic features of whiteflies in the family Aleurodidae, which are small insects crucial as vectors of plant diseases like mosaic viruses.

Background: The family Aleurodidae comprises minute insects known as whiteflies, which are economically significant as vectors of plant diseases, particularly mosaic viruses.

Key Identical Characteristics of Family Aleurodidae

1. Size: Whiteflies are minute insects, typically measuring between 1 to 3 millimeters in length. They resemble tiny moths with an opaque body.
2. Wings: Adults have two pairs of wings with reduced venation. These wings are covered by a fine whitish dust or powdery wax, giving them a white coloration.
3. Antennae: Fully developed antennae consisting of seven segments.
4. Eyes: Reniform compound eyes are present, along with two ocelli.
5. Rostrum: The rostrum is three-segmented.
6. Tarsi: Tarsi are nearly with two equal segments, each equipped with paired claws and an empodium.
7. Vasiform Orifice: A characteristic feature of the family Aleurodidae is the presence of the vasiform orifice, which is located on the dorsal surface of the last abdominal segment in both nymphs and adults. Through this orifice, whiteflies excrete honeydew.
8. Metamorphosis: Metamorphosis in whiteflies is complex. The first instar young ones are active crawlers, but subsequent immature stages are sessile and resemble scales. The scale-like covering is a waxy secretion of the insect. The wings develop internally during metamorphosis, and the early instars are called larvae. The next-to-last instar is a quiescent

stage called the pupa, during which the wings are developed internally and emerge at the final molt of the last larval instar.

9. Eggs: Whitefly eggs are very characteristic, being provided with a pedicel that sometimes exceeds the length of the egg.
10. Reproduction: Sexual reproduction is common in whiteflies, but parthenogenesis is also observed.
11. Vectors of Plant Diseases: Whiteflies are vectors of plant diseases, particularly mosaic viruses.
12. Examples:
    - Cotton Whitefly: *Bemisia tabaci*
    - Citrus Whitefly: *Dialeurodes disperses*
    - Sugarcane Whitefly: *Aleurolobes barodensis*

Activity

1. Collecting and Observing: Gather specimens of cotton whiteflies, citrus whiteflies, and sugarcane whiteflies. Observe these insects for the different key identical characters mentioned, including body size, wing characteristics, antennae structure, presence of the vasiform orifice, and other morphological features. Document these observations to facilitate accurate identification and classification of whiteflies within the family Aleurodidae.

27

Taxonomic Characteristics of Pscocptera and Malophaga

Objective: To elucidate the taxonomic characteristics of Psocoptera and Mallophaga, emphasizing their distinctive morphological features, ecological roles as parasites, and significance in entomology and veterinary sciences

Key Characteristics of Pscocptera

1. Adult usually has two pair of membranous, separately veined wings that are held roof like over the body.
2. Head has a distinct 'y' shaped epicranial suture.
3. Two or three segmented tarsi.
4. Chewing type mouthparts.
5. Forewings are larger than hind wings.
6. Dorsal pair of labial glands are modified in to silk glands.

Importance: they feed on paper paste of book binding, fragments of animal and vegetable matter and stored products. They are also damage dry preserve insects and herbarium specimens.

Key Characteristic of Malophaga

1. Chewing lice are parasite of both birds and mammals.
2. Wingless insect and have large head.
3. Body is dorsoventrally flattened.
4. Compound eyes are reduced.
5. Prothorax is invariable free and not fused with pterothorax. Meso and metathorax may be free or fused.

6. The tarsus is either unsegmented or two segmented.
7. Eggs are called nits and are cemented to the feathers.

Importance : They are obligate parasites on birds and less frequently on mammals. They severely infest the poultry bird, Affected birds will become restless and peck at one another continuously, leading to loss of plumage.

Activity

1. Draw neat labeled diagrams of psocoptera and malophaga insects with specific character

28

Comparative Taxonomic Characteristics of Orthopteroid and Hemipteroid

Objective: To differentiate and identify orthopteroid and hemipteroid groups based on their distinct taxonomic characteristics.

S.No.	Orthopteroid groups	Hemipteroid groups
1.	Nymphs with ocelli	Nymphs lack ocelli
2.	Chewing mouth parts, gular region well developed	Chewing mouthparts in primitive groups, gradual development of mouth parts.
3.	Hind wing dominant in flight	Forewing dominant in flight Reduction
4,	Antennae long, multiarticulate	in antennal segments With 3 or fewer
5.	Tarsomere number variable	tarsomeres Malphigian tubules 4 or 6
6.	Malphigian tubules numerous	number CNS strongly concentrated in
7.	CNS with separate ganglia in thorax and abdomen	thorax Herbivorous, predators, parasite, non-
8.	Herbiorous, predators, very few parasite, some social and sub-social	social, few sub social

Activity

1. Collect and identify the orthopteroid and hemipteroid insect

29

Habit, Habitat and Taxonomic Characteristics of Dermaptera

Objective: To study the habitat, habits, and key morphological characteristics of Dermaptera (earwigs).

Habit and Habitat: They are terrestrial but mostly nocturnal insects and are often attracted to light. They cannot fly (exception Labia minor). During day time, they hide away under stones, hollow stems or bark, vegetation or in the crackes of soil. They may live together in colonies . Female guards her eggs showing unique parental care. Some are parasitic (Hemimeus, Arixenia). They are omnivorous

Key Characteristics of Dermaptera

1. Head: Prognathous, clypeus is differentiated into sclerotized postclypeus and membranous anteclypeus.
2. Epicranium with Y-shaped suture.
3. Compound eyes: well developed.
4. Ocelli: Often absent or vestigial.
5. Mouthparts: Large biting-chewing type, maxillary palp-5 segmented, ligula and superlinguae bilobed.
6. Thorax: With large prothorax, metanotum fused with first abdominal segment. Metasterna with distinct apophysial pits
7. Wings Greatly reduced or absent (e.g., Anisolabis,Hemimerus, Arixenia etc. The forewings are modified into truncated, leathery, veinless, very short tegmina about half of the length of an abdomen.
8. The hindwings are membranous, semicircular, mostly formed from greatly extended anal lobe, while preanal part of wings is highly sclerotized, short and with reduced radial and cubital veins. The rest of part of wings is provided with radially arranged, secondarily developed veins. They are folded longitudinally in a fan-like fashion along with two transverse folds and at rest, are concealed under the tegmina

9. Legs: Equal in size, short, with 3-segmented tarsi.
10. Abdomen: Eleven segmented, last segment modified into an epiproct and a pair of paraprocts. First tergum is fused with the metathorax. The 8th and 9th segments are reduced or overlapped by the 7th segment, particularly in the females.
11. The Ovipositor: It is absent or reduced (Forficulina) consisting two pairs of valves.
12. The Male Genitalia: There are two penes in Pygidicranidae, Carcinophoridae and Labiduridae or only one in other families.
13. The Cerci: Modified into unjointed forceps (Forficulina), In female they are short, straight and undented while in male they are long, forked and armed (Forficula). Cerci serve defensive, prehensile and copulatory functions.
14. Four to five nymphal instars can be seen in insect in one season and show parental care for eggs

Activity

1. Collect the listed insects and observe them for different key characters(Dermaptera)

30

Habit, Habitat and Taxonomic Characteristics of Order Embioptera

Objective: To study the habitat, habits, and key morphological characteristics of Embioptera

Habit and Habitat: The Embioptera are small, fragile, soft bodied, weakly flying, brown or yellowish-brown coloured insects. They live under stones barks and avoid the light but males are generally attracted by light. They show sexual dimorphism as the the males being winged while the females remain apterous. Strikingly, they construct silken tunnels for inhabitation and can run forward as well as backward with equal efficiency. They form a colony by living in a series of superimposed tunnels communicating with each other. Males are carnivorous, while females are herbivorous. Females show parental care for their offsprings.

Key character Embioptera

1. Head: Prognathous, freely movable, with a ventral gula.
2. Compound Eyes: Elliptical in shape, smaller in female.
3. Ocelli: Absent.
4. Antennae : Filiform, short even than head.
5. Mouthparts: Well developed orthopteroid type. Mandibles are slender in male, broad in female. Maxillary palps are 5-segmented. Labial palps are 3-segmented.
6. Thorax: Prothorax is smaller than meso and metathorax. It is divided into anterior and posterior divisions by a transverse suture in male while in female it is elongate and narrow.
7. Wings: Only in males; Identical, flexible, membranous, covered with hairs; Radial vein is enclosing blood sinus: other veins are reduced, wing membrane is smoky in colour with longitudinal interveinal hyaline areas.

8. Legs: Mostly equal in length, fore and hind femora are A large due to deposition of depressor tibial muscles. Tarsi are 3-segmented,. Metatarsus of forelegs is swollen and possess the spinning gland. They run very fast in both forward and backward direction .
9. Abdomen:- 10 segmented well evident, 11 segment is represented by a pair of asymmetrical cerci . terga of 10th segment is divided into 2 asymmetrical plates or hemitergites in the males.
10. Cerci: 2 segmented, asymmetrical in males .
11. Tracheal system is 2 thorasic and 8 abdomenal spiracles
12. 4 nymphal instars are found in these insect order.

Activity

1. Collect the listed insects and observe them for different key characters (Dermaptera)

31

Taxonomic Characteristics of Arctiidae and Acrididae Family in Insect

Objective: To differentiate and document the key morphological characteristics of insects from the families Arctiidae (tiger moths) and Acrididae (grasshoppers).

Arctiidae	**Acrididae**
It comes under order Lepidoptera Metamorphosis is complete Common name is tiger moth Wings are conspicuously spotted or banded They are nocturnal and attracted to light Larvae is either sparsely hairy or densly haired (wooly bear) Example: *Amscata moorei* (red hairy caterpillar)	It comes under orthoptera Incomplete metamorphosis Common name is grasshopper Tarsi is 3 segmented Ovipositor is short and horny Tympanum is located on one or either side of the first abdominal segments Larvae resemble adults morphology except for the genital organs Example : rice grasshopper (*Heigroglephus banian*)

Activity

1. Collect and observe the morphometric character of red hairy caterpillar and grasshopper

32

Taxonomic Characteristics Endopterygota Superorder and Their Present Status with Respect to Sister Groups

Objective: To identify and understand the taxonomic status and relationships of orders within Endopterygota, focusing on their distinct larval and adult stages and current sister group hypotheses.

1. Endopterygota comprise insects with immature (larval) instars that are very different from their respective adults.
2. The adult wings and genitalia are internalized in their preadult expression, developing in imaginal discs that are evaginated at the penultimate molt. Larvae lack true ocelli. The "resting stage" or pupa is nonfeeding and precedes an often active pharate ("cloaked" in pupal cuticle) adult.
3. Two or three groups currently are proposed among the endopterygotes, of which one of the strongest is a sister group relationship termed amphiesmenoptera between the trichoptera (caddisflies) and Lepidoptera (butterflies and moths).
4. A plausible scenario of an ancestral amphiesmenopteran taxon has a larva living in damp soil among liverworts and mosses followed by radiation into water (trichoptera) or into terrestriality and phytophagy (lepidoptera).
5. A second (usually) strongly supported relationship is between order neuroptera and sister group coleoptera.
6. A third, postulated relationship antliophora, unites diptera (true flies), siphonaptera(fleas), and mecoptera (scorpionflies and hanging flies).
7. Debate continues about the relationships of these taxa, particularly concerning the relationships of siphonaptera.

8. Fleas were considered sister to diptera, but molecular and novel anatomical evidence increasingly points to a relationship with a curious looking mecopteran.
9. Strepsiptera is phylogenetically enigmatic, but resemblance of their first-instar larvae (called triungulins) to certain coleoptera, notably parasitic rhipiphoridae, and some wing base features have been cited as indicative of relationship.
10. This placement is becoming less likely, as molecular evidence (and haltere development) suggests a link between Strepsiptera and Diptera.
11. The long-isolated evolution of the genome can create a problem known as "long- branch attraction", in which nucleotide sequences may converge by chance events alone with those of an unrelated taxon with a similarly long evolution, for the strepsipteran notably with diptera. The issue remains unresolved.

Several positions have been proposed for coleoptera but current evidence derived from female genitalia and ambivalent evidence from eye structure supports a sister group relationship to neuropterida.

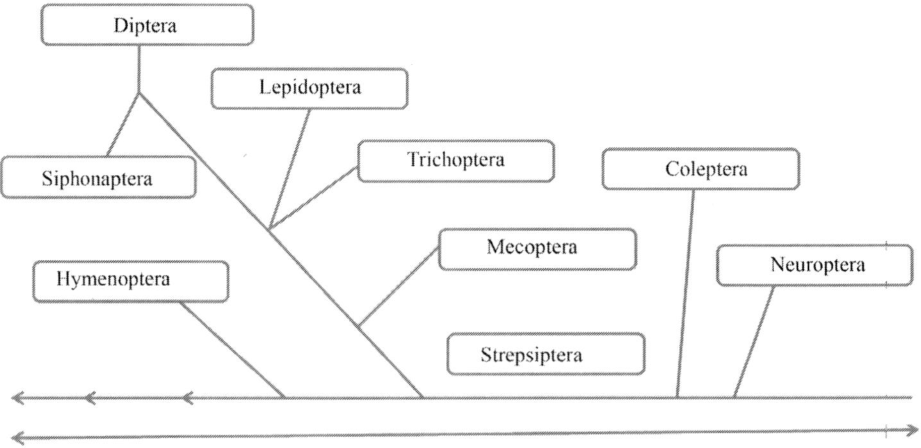

Sister group of endopterygote

Activity

1. Collect and identify the endopterygote insect with sister group

33

Taxonomic Characteristics of Order Strepsiptera and Mecoptera

Objective: To explore and document the taxonomic characteristics of Strepsiptera and Mecoptera, highlighting their unique morphological features and biological adaptations.

Taxonomic character Strepsiptera

2. Strepsiptera form an enigmatic order of nearly 400 species of highly modified endoparasites, most commonly of Hemiptera and Hymenoptera, and show extreme sexual dimorphism.
3. The male has a large head with bulging eyes comprising few large facets and lacks ocelli.
4. The antennae are flabellate or branched, with 4 to 7 segments.
5. The forewings are stubby and lack veins, whereas the hind wings are broadly fanshaped, with few radiating veins; the legs lack trochanters and often also claws.
6. Females are either coccoid-like or larvi form, wingless, and usually retained in a pharate (cloaked) state, protruding from the host.
7. The first instar is a triungulin, without antennae and mandibles, but with three pairs of thoracic legs; subsequent instars are maggot-like, lacking mouthparts or appendages.

Taxonomic character Mecoptera

1. Mecopterans are holometabolous insects comprising about 500 known species in nine families, with common names associated with the two largest families Bittacidae (hanging flies) and Panorpidae (scorpion flies).
2. Adults have an elongate ventrally projecting rostrum, containing elongate, slendermandibles and maxillae, and an elongate labium.

3. The eyes are large and separated, the antennae filiform and multi segmented.
4. The fore- and hind wings are narrow, similar in size, shape, and venation, but often are reduced or absent.
5. The legs may be modified for predation.
6. Larvae have a heavily sclerotized head capsule, are mandibulate, and may have compound eyes comprising 3 to 30 ocelli.
7. The thoracic segments are about equal and have short thoracic legs with fused tibia and tarsus and a single claw. The pupa is immobile, mandibulate, and with appendages free.

Activity

1. Collect and observe the morphometric character of strepsipteran and mecopteran

34

Taxonomic Characteristics of Order Siphonptera and Trichoptera

Objective: To study and document the taxonomic characteristics of Siphonaptera and Trichoptera, emphasizing their unique morphological features and ecological adaptations.

Taxonomic characteristics siphonaptera

Siphonaptera is a highly modified order of holometabolous insects, comprising some 2400 species, all of which are bilaterally compressed, apterous ectoparasites.

1. The mouthparts are specialized for piercing and sucking, lack mandibles, but have an upaired labral stylet and two elongate serrate, lacinial stylets that together lie within a maxillary sheath.
2. A salivary pump injects saliva into the wound, and cibarial and pharyngeal pumps suck up the blood meal.
3. Fleas lack compound eyes and the antennae lie in deep lateral grooves.
4. The metathorax houses very large muscles associated with the long and strong hind legs which are used to power the prodigious leaps made by these insects

Taxonomic characteristics trichoptera

1. Trichoptera contains about 45 extant families containing some 10,000 described species, with estimates of undescribed (mostly Southeast Asian) species diversity some four to fivefold higher.
2. Trichoptera are holometabolous; the moth like adult has reduced mouthparts lacking any proboscis, but with three- to five segmented maxillary palps and three segmented labial palps.
3. The antennae are multi segmented and filiform and often as long as the wings. The compound eyes are large, and there are two or three ocelli.

4. The wings are haired or less often scaled and differentiated from all but the most basal Lepidoptera by the looped anal veins in the forewing and absence of a discal cell.
5. The larva is aquatic, has fully developed mouthparts, has three pairs of thoracic legs (each with at least five segments), and lacks the ventral prolegs characteristic of lepidopteran larvae.
6. The abdomen terminates in hook bearing prolegs.
7. The pupa is also aquatic, enclosed in a retreat often made of silk, and possesses functional mandibles to aid in emergence from the sealed case.

Activity

1. Collect and observe the taxonomic character of siphonopteran and trichopteran insect

35

Identify Agriculturally Important Insects Order with Some Identical Characters

Objective: To identify agriculturally important insect orders and understand their key morphological characteristics relevant to their impact on agriculture.

Agriculturally important insect orders

1. **Lepidoptera:** This includes all moths and butterflies, characterized by scaly wings that are either darkly coloured or brightly coloured respectively catterpillars are the damaging stage
2. **Diptera:** Includes all the true flies .appears to have a single pair of wings because hind wings are very much redused to a balancing organ called halters . maggots (larval stage) are the damaging stages. Consume liquids only through sucking mouthparts. Complete metamorphosis - a change in physical form from earlier stages in life to adulthood, includes 4 stages.
3. **Coleoptera:** Includes all the beetles and weevils characterized by the presence of heavily sclerotized fore wings boths grubs and adults are the damaging stages undergo complete metamorphosis
4. **Orthoptera:** Includes all thegrasshoppers .theymaybe either solitary or in gregarious form. usually short horn grasshoppers turns to gregarious form commonly called locusts. generally cylindrical body, with elongated hindlegs and musculature adapted for jumping. They have mandibulate mouthparts for biting and chewing and large compound eyes, and may or may not have ocelli, depending on the species.
5. **Thysanoptera:** Includesthripssoft bodied minute insect barely visible to naked eyes, some species may attain a length of 12 mm, causes curling of leaves with a streaks on the leaf surface larval and adult stages are damaging stages. The compound eyes usually have large facets, ocelli are present in winged adults only. Another curious feature of thrips is that the right mandible is absent.

6. Isoptera: Includes termites.he compound eyes are frequently reduced, the antennae are long and multisegmented, and the fore wings and hind wings are generally similar, membranous, and have restricted venation they resemble ants in
7. their structure they are soft bodied insects and prefer dry humid conditions their damage is more prevalent in unmanaged farms
8. Neuropteran: Includes green lace wing bugs .they are beneficial insects found to be a predator of many crop damaging insects Neuropterans are soft-bodied insects with relatively few specialized features. They have large lateral compound eyes, and may or may not also have ocelli. Their mouthparts have strong mandibles suitable for chewing, and lack the various adaptations found in most other endopterygote insect groups
9. Hemiptera: Includes all the bugs, aphids, whiteflies, leafhoppers. they are the sucking pest . sucks the sap from plant surface and cause stunting of plant. The true bugs have forewings that are hardened at the base and membranous at the tips. They sit flat over the abdomen hiding the membranous hind wings. The head and proboscis can flex forward. The hoppers have forewings that are uniform in texture and are held like a tent over the abdomen
10. Hymenoptera: Includes insets like wasps bees and sawflies .most of them are beneficial they serve as parasitoids . some of the insects like honeybees helps in production of products which are utilized by humans since ages. Hymenopterans are chiefly small to medium-sized insects, usually with four membranous wings and a narrow waist that sets off the abdomen from the thorax, or middle region of the body. The mouthparts may be either of the biting type or of the chewing and lapping type.

Activity

1. Collect and identify the agriculturally important insect on the basis of taxonomic character

Index